Man and Forest
No. 1

General Editors
Klaus Seeland
Franz Schmithüsen

A project of

Chair of Forest Policy and Forest Economics
Swiss Federal Institute of Technology (ETH)

and

The Deutsche Gesellschaft für
Technische Zusammenarbeit (GTZ)

Man and Forest

***General Editors*: Klaus Seeland; Franz Schmithüsen**

The Man and Forest Series consists of monographs and readers on the use and management of forests and an account of forest life styles in various regions of South Asia. It aims at providing the findings of an eight years research project (1992-1999) to the public. The volumes in this series are the outcome of a cross-cultural research network between the Chair of Forest Policy and Forest Economics, Swiss Federal Institute of Technology (ETH), the Deutsche Gesellschaft für Technische Zusammenarbeit (GTZ) and Non-Government as well as Government Organizations in India, Nepal and Bhutan. The series welcomes contributions covering relevant issues in the man-forest interface

MAN IN THE FOREST

MAN IN THE FOREST

Local Knowledge and Sustainable Management of Forests and Natural Resources in Tribal Communities in India

Editors

Klaus Seeland
Franz Schmithüsen

D.K. Printworld (P) Ltd.
NEW DELHI - 110 015

Cataloging in Publication Data — DK

Man in the forest: local knowledge and sustainable management of forests and natural resources in tribal communities in India / editors, Klaus Seeland, Franz Schmithüsen.

(Man and forest; no.1).

Includes bibliographical references.

Includes index.

1. Sustainable forestry -- India. 2. Conservation of natural resources -- India. 3. Forests and forestry -- Economic aspects -- India. I. Seeland, Klaus. II. Schmithüsen, Franz. III. Series: Man and forest; no. 1.

ISBN 81-246-0152-6

First Published in India in 2000

Supported by: Flanking Program for Tropical Ecology, Deutsche Gesellschaft für Technische Zusammenarbeit (GTZ)

Published and printed by:

D.K. Printworld (P) Ltd.

Regd. office : *'Sri Kunj'*, F-52, Bali Nagar

New Delhi-110015

Phones : (011) 545-3975, 546-6019; *Fax* : (011) 546-5926

E-mail : dkprint@4mis.com

Preface

In 1991, a research cooperation was established between the Chair of Forest Policy and Forest Economics at the Swiss Federal Institute of Technology and Indian NGOs. In a joint research project called "Indigenous Knowledge, Environmental Perception and Traditional Management Systems of Forest Dwelling Tribes and Non-Tribal Village Communities in Selected Regions of Orissa/ India" that was conducted during 1992-95 cross-cultural and interdisciplinary investigations tried to get access to the tribal forest world of Orissa. This project produced a vast amount of data covering a broad scope of the use and management of trees and forests which were presented at an international seminar held in New Delhi at Max Mueller Bhavan in March 1995. One of its significant positive results was the decision to embark on a follow-up and establish a research network in which trees, forests and forest management in the wider Himalayan context are investigated and research partners from India, Nepal and Bhutan are included.

One of our partners, with whom we cooperated from the very beginning, was the Council of Cultural Growth and Cultural Relations in Cuttack, Orissa and in particular its General Secretary, Mr. G. Bh. Patnaik who was the initiator and facilitator of many valuable contacts at every level. Research, we dare say, would not have been possible without his enthusiastic participation particularly in the take-off phase of the project (1991-93). His wife, Prof. Dr. (Mrs.) K. K. Patnaik, Dr. H.C. Das and later on Dr. J. Dash were advisors and guides to the research scholars, Dr. M. K. Jena, Ms. P. Pathi, S. Behera, Dr. Savyasachi and Ms. A. Mishra, who worked for the project from a period of some months to seven years. Other partners were A. S. Mehta, Secretary of Seva Mandir, a NGO located in Udaipur and L.P. Bharara, principal scientist (now retd.), from the Central Arid Zone Research Institute, Jodhpur, Rajasthan, who either conducted research on their own or integrated indigenous knowledge into their practical development work. In the second phase Dr. N. Patnaik, Director of the Social Science and Development Institute in Bhubaneshvar, Orissa, joined the project. For the international seminar he had already been invited as an expert on tribal affairs. All of them inspired the discussions in the research network and thus contributed to the success of the project despite the many difficulties and obstacles that sometimes made fieldwork tiresome.

The main objective of the second phase of this project network (1996-99) is to investigate indigenous knowledge of trees and forests among various forest dwelling tribes as a mode of appropriation of nature in a wider Himalayan perspective. In this phase SAGUN, a Nepalese NGO that has been working on indigenous knowledge and forest related matters with other partners, and government officials from Bhutan who do research on forest and development policy have joined the research network. In all empirical studies which have been conducted within the network, an emphasis is laid on local, regional and national natural resource policies. As a cross-cultural cooperation, scientific knowledge and development practice at the local level were to be seen as interlinked and theoretical and methodological support for the local researchers were provided by the Chair of Forest Policy and Forest Economics at the Swiss Federal Institute of Technology in Zurich, Switzerland. The German Ministry for Economic Cooperation has been financing the major part of the project through the German Technical Cooperation (GTZ) within the scope of promoting its programme of tropical forest research, now merged with the Ecological Flanking Programme for the Tropics (Toeb). In this interface the research project is to promote local, regional and international cooperation and exchange among researchers, an extended web of partner-NGOs and university institutions with environmental and socio-cultural concern in the wider Himalayan region.

We wish to thank our research partners in the network who have cooperated with us and have been exchanging their data, experiences and views among each other in a fruitful way as well as the financing agencies for their generous assistance. A rewarding cooperation between comparatively small and rather independent research units located in two continents and with different professional backgrounds and experiences cannot be taken for granted. What helped us to overcome the difficulties the project had to face, was the shared interest in the many ways in which some of India's rural communities perceive forests, the readiness to work as equal partners, and the personal commitment of the researchers involved.

The highly welcome contributions of S. Brodt, R. Pant, S. Vasan and S. Kant supplement the papers presented at the international seminar, adding more to the scope and regional distribution of research on indigenous knowledge of forests in India.

Many thanks go to those who were involved in administrative affairs and facilitated the research project and the international seminar in New Delhi, particularly the head office and the Project Administration Service of the GTZ in India and the Max Mueller Bhavan in New Delhi for their very successful hosting of the seminar and their assistance in so many practical matters which contributed to make it a real success.

The editors express their gratitude to all who have assisted in preparing and editing this volume and to Mrs. Nadja Ottiger in particular whose help in finalising the manuscripts was most valuable.

Klaus Seeland
Franz Schmithüsen

Contents

Contributors
(in order of their contributions)

Prof. Dr. Franz Schmithüsen
Project Leader of the Research Project "Indigenous Knowledge on Forests"

Chair of Forest Policy and Forest Economics, Swiss Federal Institute of Technology, Zurich, Switzerland

PD Dr. Klaus Seeland
Project Leader of the Research Project "Indigenous Knowledge on Forests"
Chair of Forest Policy and Forest Economics, Swiss Federal Institute of Technology, Zurich, Switzerland

Dr. Sonja Brodt
Research Intern, East-West Center Honolulu, Hawaii, USA

Dr. Nityananda Patnaik
Director of the Social Science and Development Research Institute, Bhubaneshvar, India

Dr. Laj Pal Bharara
Principal Scientist (retd.), Central Arid Zone Research Institute (CAZRI), Jodhpur, India

Dr. Ruchi Pant
Director of the Eastern Himalayan Programme of the Ashoka Trust for Research in Ecology and the Environment (ATREE), India

Dr. N.P. Melkania
Professor of Applied Science Forestry NERIST/NIRJULI
Itanagar, Arunachal Pradesh, India

Sudha Vasan
Doctoral Candidate, Yale School of Forestry and Environmental Studies, New Haven, USA

Dr. Hari Chandra Das
Director (retd.) of the State Museum of Orissa, Bhubaneshvar, India

Dr. Shashi Kant
Member of the Faculty of Forestry, University of Toronto, Canada

Dr. Mihir Kumar Jena
Research Scholar of the Research Project "Indigenous Knowledge on Forest"
Council of Cultural Growth and Cultural Relations, Cuttack, India

Sricharan Behera
Research Scholar of the Research Project "Indigenous Knowledge on Forest"
Council of Cultural Growth and Cultural Relations, Cuttack, India

Padmini Pathi
Research Scholar of the Research Project "Indigenous Knowledge on Forest"
Council of Cultural Growth and Cultural Relations, Cuttack, India

Prof. Dr. Kamala Kumari Patnaik
Co-Director of the Research Project "Indigenous Knowledge on Forest"
Council of Cultural Growth and Cultural Relations, Cuttack, India

Dr. Jan Brouwer
Director of the Centre for Advanced Research of Indigenous Knowledge Systems, Mysore, India

1

The Meaning of Indigenous Knowledge in the Use and Management of Renewable Natural Resources

Examples from applied research in India

Klaus Seeland
Franz Schmithüsen

WHENEVER man, following developments in civilisation, was concerned with the management of renewable natural resources, forests definitely played a vital role in this process (Harrison 1992). Nowadays, the economically developed world as well as some of the underdeveloped economies still depend on their own and the world's forests for their survival, yet in different ways and meanings (Seeland 1997). Conservation and sustainable management have become central issues for almost all societies in order to either meet shortages in timber or non-timber-forest-produce supply or face the challenges of environmental pollution. Economic development in times of increasing environmental hazards tends to create a dilemma in many societies which is bound to grow despite the often narrow limitations of a declining resource base.

Economic growth, scarcity of resources and reckless pollution demand a well balanced socio-cultural concept to overcome the inherent dilemma in these three dimensions of development. Where economic and political measures fail are partly withdrawn in a process of deregulation, socio-cultural traditions such as indigenous knowledge of resource use and management have inevitably to fill the gap (Warren/Slikkerveer/Brokensha 1995). First of all, political self-reliance has to be based on a responsible and sustainable management of renewable natural resources. Only then, a solid social development based on equity and a rationale of cooperation, as far as the use and management of renewable natural resources are concerned, will be possible (Woodley 1991).

Indigenous knowledge of forests and natural resources has been the subject of research for more than two decades now. The owing to results that have been obtained are encouraging, yet not always satisfactory problems in the implementation of this knowledge into practical development work. One possible reason for this is the fact that indigenous knowledge is more a life-style than merely a set of traditional skills and know-how which may easily be transferred to other socio-cultural settings. However, the appeal of local and indigenous knowledge is not its transfer or operationalisation, but its local value as a means of cultural persistence and continuity (Wiersum 1997). Self-esteem and the empowerment of local socio-economic and political structures will ultimately depend upon the civilisational value indigenous or local knowledge has in a phase of rapid globalisation. It is particularly valuable wherever social development on the basis of cultural identity is desirable or favoured. Political structures and hierarchies of power connected with social imbalances and privileges of the knowledgeable and traditional elites may also coincide with local and indigenous knowledge and traditional technologies. This political ambiguity of traditional knowledge structures, however, and their manifestation in material culture and social stratification is part and parcel of their socio-political impact and its inherent dynamics.

The contributors to this volume are of various cultural backgrounds and represent different generations, kind of professional training and experiences. Their contributions show the wide-ranging value of indigenous knowledge in coping with problems of environment and development in the twenty-first century. Some contributions refer to a not-so-distant past when most of the tribes of India lived an almost independent and self-reliant life in the forests. This fact seems to a certain extent to have persisted in the minds of some of the contributors, although the tribals' forest habitat has undergone tremendous changes both in quantity and quality over the last 50 years. To either glorify or criticise the tribals' life-style does not do justice to them. They are claimed to be responsible for considerable damage to the forests due to practising of slash-and-burn agriculture, and their excessive hunting and gathering practices are said to threaten endangered species and cause a remarkable loss in biodiversity. As with any other society which has to face the development process of modernisation and the social changes resulting from it, the tribal population of India had to look for options and possibilities to survive in an emerging nation-state with all its problems of demands for resource use equity, a balanced growth and resource supply for all sections of the society. The tribal-non-tribal interface characterises rural life in many remote areas of India today. Paternalism from the side of the Hindu mainstream society, conflict and a rapid change in authentic tribal lifestyle dominate the Hindu-tribal relationship, yet there are many social activists who selflessly work for improvements for these marginalised forest dwellers, where governmental aid-programmes such as the so-called micro plans, fail (Fernandes/Menon/Viegas 1988). Some contributors to this book emphasise the need for an intermediate technology to be derived out of age-old local and socially appropriate, as well as environmentally sustainable, management practices. Their plea for the survival of marginal man and the abundant nature in the tropics and subtropics, and the interest of those researchers who are interested in ethno-biology and sociology for the sake of recording indigenous lifestyles, meet with a common concern for conservation and alternatives to modern mainstream development.

The contributions here are a collection of research findings presenting accounts of experience and empirical evidence in the fields of forest management, social anthropology, ethno-botany, economy, forest policy and cultural history. Except four contributions, all of them were read and discussed at the international seminar "Indigenous Knowledge On Forests" in New Delhi in 1995. The original contributions to the seminar have already been published in 1997 in the series "Forstwissenschaftliche Beiträge, No. 19 of the Chair of Forest Policy and Forest Economics, Eidgenössische Technische Hochschule Zürich" in Switzerland. The supplementary papers add valuable findings, insights and information to the existing corpus of papers from the seminar. Other contributions, mostly shorter ones, partly being policy statements or specialised accounts which transcended the scope of the general topics raised, have been omitted. Together, all contributions reflect the state of the art within a range of selected scientific disciplines which inspire the policy dialogue taking into account the ascribed benefits, as well as the limitations, of indigenous knowledge in general, and with regard to forest management by local tribal people in particular.

It is the objective of this selection of papers to show the potentials of as well as the socio-cultural and political obstacles to protecting and managing the forests of the Indian subcontinent in a sustainable way. Traditional modes of production and conservation are considered as relevant for future cultural and economic development, and this point of view is reflected in most of the papers. The general aim of indigenous knowledge as a scientific topic is the search for appropriate means to meet the challenges and deficiencies of modernity, i.e., the rising number of poor people living in precarious environments with very limited prospects for self-sufficiency in food production or the sustainable management of their local resources.

F. Schmithüsen highlights the socio-cultural legacy and the political development, importance and meaning of indigenous knowledge research for sustainable forestry practices. Thus he gives the framework in which the regional examples, covering North (Himachal Pradesh, Arunachal Pradesh, Uttar Pradesh),

West (Rajasthan), East (Orissa), South (Karnataka) and Central India (Madhya Pradesh), emphasise the social meaning of forests as a cultural legacy. K. Seeland's contribution gives a theoretical introduction to various aspects of indigenous knowledge, hints at the historical development of its background and draws a sketch of its practical relevance for development cooperation. S. Brodt's is the first contribution of a regional case study. In her example located near the city of Bhopal in Madhya Pradesh she describes the nexus between modern, i.e., global and scientific knowledge and local techniques in agriculture and tree cultivation. She shows that social status, caste, prestige, economic wealth, education, structures of communication and age matter most in the application of knowledge and innovation of new cultivation practices. The complexity of local knowledge and differences in access to global technical know-how has, until now, led to its preservation of, or at least to forms of coexistence between, indigenous and other systems of knowledge. Nityananda Patnaik gives an account of the broad eco-diversity of the forest world of Orissa and its reflection in the tribal people's lifestyle. He stresses the view that there is a symbiosis of forest dwelling tribes and their surroundings and every fabric of their social and cultural life seems to be interwoven with so many aspects of the forest environment. Indigenous knowledge of land resource conservation in Rajasthan and its relation to the local people's perception is the focus of L. P. Bharara's contribution. Soil conservation, traditional methods of water harvesting and other ways and means to adjust to the hazards of drought and a poor and fragile environment characterise the livelihood pattern of the desert dwellers in the semi-arid and arid zones of Rajasthan. In his view, the scarce resource base for survival in desert areas has ever since been endangered and has recently come even more under pressure due to the rising number of livestock and human population. It seems that neither the state nor cooperation from outside can help to solve the problem; the most appropriate and sustainable option is to enhance self-reliance by using indigenous knowledge to strengthen the local adjustment capacities to scarcity and environmental hazards. R. Pant shows in her paper on forest resources management in the two traditional mountain communities of the Apatani and Nishi tribes in the North-Eastern Himalayas of Arunachal Pradesh that conflicts over resource use and their resolution was and still is to a large extent most efficiently tackled by local institutions. The functioning of some institutions relevant for resource management is described and several case studies show how successful they are as compared to the respective administrative services of the state. The contribution raises the question of how appropriate formal legislation in general and forest legislation in particular is to solve conflicts in a tribal-non-tribal social and administrative interface. The Central Himalaya with examples of traditional modes of forest and tree management on the basis of indigenous knowledge is the topic treated by N. P. Melkania. They have been developed by the farmers over time and are adjusted to specific ecological and ethno-economic conditions. In the author's view, development of forest management strategies and plantation projects in the region must proceed from local strategies and reinforce indigenous knowledge rather than try to impose new and often alien ideas on the farmers. Suggestions for sustainable management of tree-based natural resource management are outlined for the Central Himalayan region of Uttar Pradesh and the Himalayan ecosystem as a whole. The local meaning of traditional herbal medicine as an example of indigenous knowledge in the Sainj valley of the Western Himalayas is investigated by S. Vasan. In several spheres of occupation such as transhumance, the management of a sacred grove, or collecting herbs and medicinal plants, she shows adaptations to the conditions and requirements of emerging modern demands and markets in this remote part of Himachal Pradesh. These changes have mostly resulted in a revitalisation and re-valuation of indigenous knowledge and an increasing involvement and participation of local communities. The contribution of H. C. Das is an inquiry into the religious history and art with particular reference to some selected plant and tree species. It provides an introspection into the relationship of man and forest from the ethno-historical perspective. From prehistoric times certain plants and trees have been deified and they continue to play a vital role in Hindu religion and Indian civilisation in general. The salient features of tribal economy in five selected villages of Raisen District of Madya Pradesh are analysed by S. Kant. It is clear from this interesting analysis that the contribution of certain forest products to the household income, the seasonal importance of forest produce and the rates of return on labour enhance the capacity of tribal forest dwellers to survive in a marginal forest environment. By making use of forest produce they are in a position to compensate for the income inequalities due to seasonal imbalances of agricultural yields and social inequalities. In this way, forest products contribute to improved food security in years of calamities and mitigate uncertainties in the income structure. A comparison of three tribal communities of Orissa, the Hill Kharia, Kuttia Kondh and Saora, to investigate their indigenous ideas of protection of plant species on religious grounds is presented by M. K. Jena. This paper shows how the sacred world-view of these tribal groups plays a role in the preservation and conservation of forests. Sacred complexes in forests matter to all of these communities as far as mythical, economic and ecological aspects and their role for the preservation of biodiversity are concerned. S. Behera provides the reader with a very complex and enlightening account of how the Saora tribal community of Orissa perceives resources in the material, social and spiritual dimensions of their life-world. The most intriguing aspect of this contribution is that social resources are made part of the natural resources. Social resources are meant to enhance the social cohesion, the equity among fellow villagers and the co-operation and solidarity among the members of the tribal community. It can be characterised as the process of societal appropriation of natural resources. They are classified according to their logic of being linked to each other, to the extent they matter in the institutional set-up of the community and their contribution to the livelihood system. An account of the holistic view of the Saora cosmology and its symbolic representation in art is given in P. Pathi's contribution.

The Saora world is run by spiritual powers which act at several levels and their environmental perception can only be understood with regard to the interaction of these different layers of spiritual and material reality. Their iconography is a key to decipher their complex and manifold cosmology and can be taken as a prerequisite to obtain access to their indigenous environmental knowledge. The role of plants in the folk medicines of the Kuttia Kondh and Saora of Orissa is investigated in the paper of K.K. Patnaik. This very instructive article shows a lot of interesting applications of plants as remedies for various purposes. It is the result of intensive field research aiming to preserve indigenous medicinal knowledge which is endangered due to the spreading of modern medical services. This contribution adopts a pragmatic approach to reflect local indigenous knowledge in a more practical than a science-oriented way. The last paper by J. Brouwer examines south Indian artisans' views on their principal raw materials from the vantage point of the craftsman's perception. Indigenous knowledge is investigated in the context of society and culture of the Visvakarma caste of Karnataka. Although south Indian artisans consider their fauna and flora as separate from the self and society, their perceptions of their working materials — wood and stone — show a close interrelationship among social, religious, ecological and technological domains (Brouwer 1995).

Indigenous knowledge in a modern world under conditions of economic underdevelopment and precarious environments has to face criticism from inside and outside developing societies. The political implications of favouring local, decentralised socio-political structures raise suspicion at the central government level. Local empowerment and decentralisation is often claimed to strengthen the establishment of democratic decision-making processes and indigenous knowledge could help local structures to attain self-reliance in economic and thus to a certain extent in political matters as well. But encouraging revitalisation of indigenous knowledge in the context of development is ambivalent as it is always an intervention into a living social organism. International development or conservation agencies tend to apply their theoretical concepts and political programmes at the national, regional or local project level and in this process often transgress their cultural and political limitations, even if they are claimed to be on the global agenda. Whenever sections of rural societies choose to revalidate their indigenous knowledge it is a local political decision, but in its implications it is a global approach, initiated from outside and to be seen in a larger context of international and cross-cultural environmental policy. Indigenous knowledge as a social development concept is an option which can make sense and can be appropriate under certain conditions at a particular stage of crisis within the larger framework of the development of civilisation based on equity, sustainability and self-determination.

Some of the contributions to this book may reflect these aspects more or less explicitly; others may be more detail-oriented and convinced that indigenous knowledge is beyond any doubt the solution to many social and environmental problems. Being the editors of this volume, we are tempted to say that the overall value of these contributions, first of all, is that they provide the reader with empirical research findings from a regional entity, the Indian subcontinent and thus inspire the general discussion on indigenous knowledge at a culturally distinct and very practical level.

References

Brouwer, J., The Makers of the World. *Castes, Craft and Mind of South Indian Artisans*, New Delhi: Oxford University Press, 1995.

Fernandes, W., Menon, G., Viegas, Ph., Forests, Environment and Tribal Economy. *Deforestation, Impoverishment and Marginalisation in Orissa*, New Delhi: Indian Social Institute, 1988.

Harrison, R. P., *Forests, The Shadow of Civilization*, Chicago / London: The University of Chicago Press, 1992.

Seeland, K. (ed.), Nature is Culture. *Indigenous Knowledge and Socio-cultural Aspects of Trees and Forests in Non-European Cultures*, London: Intermediate Technology Publications, 1997.

Warren, D. M., Slikkerveer, L.J., Brokensha, D. (eds.), The Cultural Dimension of Development. *Indigenous Knowledge Systems*, London: Intermediate Technology Publications, 1995.

Wiersum, K. F., Indigenous exploitation and management of tropical forest resources: an evolutionary continuum in forest-people interactions. *Agriculture, Ecosystems and Environment* 63 (1997):1-16.

Woodley, Ellen, Indigenous Ecological Knowledge Systems and Development. *Agriculture and Human Values* (winter-spring) 1991:173-8.

2

The Socio-Cultural and Political Context of Sustainable Forestry Practices[1]

Franz Schmithüsen

Sustainable forestry practices as socio-cultural phenomena

FORESTS show great variety, represent an abundance of life, contain many possibilities for living beings and provide habitat for many forms of plants and animals. Abundance, growth and continuity are prominent features in man's perception of a forest. There are several definitions of forests and forest lands due to different assessment and inventory techniques. The most consistent data, especially for time comparisons, are those of the Food and Agriculture Organisation, FAO. According to this source forests and woodlands cover slightly more than 30 per cent of the world land area. About half of the world's forests and woodlands is situated in the boreal and temperate zones, whereas the other half is in the tropics and subtropics.

Human activities influence forests although their impact is difficult to assess. Some of the changes are immediate and occur in the short term. Others, and often the more important ones, are indirect and are to be understood in historical dimensions. The spatial distribution of forests and the degree of their transformation by man are the result of physical conditions

---

1. Based on a contribution published in "L'uomo e la Foresta"; Le Monnier, Firenze, 1996.

and of varying cultural patterns. This applies not only to forests that have been intensively used over centuries, but also to forests which still appear to be in a natural state. They may have been spared from human intervention due to physical inaccessibility and economic considerations, or because they represent particular spiritual values of a society. The differences between intensively used forests and those showing little or no trace of human interference reflect social rules, economic options and political decisions. In this respect all forests, including those which we still consider as natural, are cultural phenomena.

Forests and woodlands are environments which societies interpret as local surroundings, develop as resources for commercial and industrial purposes and consider as a representation of nature. In such a perspective the question is whether one should continue to perceive forests as rural or perceive them as urban in societies in which spaces form an entity and represent distinct but complementary potentialities. The varying importance of trees and forests in different cultures is the result of physical and material, socio-cultural and spiritual configurations. Physical and material conditions are the existence and potentiality of forests as a critical component of man's environment, the variety of plant and animal communities and the capacity of regeneration and self-regulation. Pertinent socio-cultural considerations are present and anticipate human needs and values, prevailing and potential uses as related to the development of technology and lifestyles, entitlements to the access of resources and public regulations on sustainable management. Spiritual considerations as apparent in myths, religion and personal notions show the place of trees and forests in man's vision of his world and of himself.

The meaning of a forest is reflected in its cultural perception and societies have developed their notions of what forests mean to them over the ages (Harrison 1992). Whatever the forests represents for the members of one society, they certainly have a different relevance for others who see and understand them within the dimensions of their own culture. The forests of today show how people have been, and still are, dependent on them, and how they make use of and interpret their environment in terms of survival and social advancement. The transformation of forest vegetation that we observe, indicates specific social needs, cultural values, and changing economic and technological processes. Forests represent a legacy and they are a testimony of the evolution of societies and their respective perceptions of nature.

Public institutions, land management practices and the performance of a society in developing sustainable use patterns are evolutionary processes (Firey 1960). The analyses of such processes reveal not only the reasons for protection and management, but also the causes and rationales for the disappearance of forest vegetation as well as for the creation of new forests. Historical developments show changes in the importance of forests in different societies, varying perceptions of forests as cultural phenomena and evolving social attitudes in dealing with perceived problems and opportunities. From a political perspective, fundamental issues which have been important during the course of time and which are relevant today are appropriation patterns, tenurial systems and changes in land use. Institutional solutions have been established in order to find a socially acceptable balance between conflicting private and collective demands, between local and national resource appropriation, and between benefits accruing to different user groups and their responsibility in protecting forests (FAO 1994).

Forests mean different things to different people. If one would ask people individually what aspects of forests are important to them, which is rarely done, we would get answers reflecting contradictory interests and convictions. The inhabitants of large cities would perhaps stress the importance of forests for leisure and recreation, people in mountainous areas might value them for the protection they give and as an attraction to tourists, and farmers might refer to the income generated from forestry activities. Some people would stress the uniqueness and beauty of the forests and the need to preserve them, and others would insist on economic benefits and employment opportunities which result from their use. Evolving attitudes and perceptions, changing opportunities and specific user groups determine the political importance of forests. To local communities they represent opportunities of many kinds and values. At the national level, commercial wood production and the development of a forest industry sector are important. In a wider perspective they are perceived as environment and wilderness.

Forests as local surroundings

The meaning of forests as local surroundings reflects many different situations (Gregersen *et al.* 1989). Communities may actually live in forest environments which provide a livelihood and represent an element of their cultural identity. In agrarian and pastoral societies, forests are part of a continuum of complementary and often interchangeable uses. The management of local resources is largely based on rotational systems with alternating phases of agriculture and forest re-growth on the same patch of land or on combined agroforestry and silvipastoral practices. Within the various production systems, trees and different stages of forests are dynamic elements of a common and continuously changing resource potential.

For people in societies with traditional land management systems, the answer to the question, what are forests and where do they begin and end, may be difficult. They would wonder whether this is a meaningful distinction. Is the area covered with trees, in which their animals graze, a forest, a pasture, perhaps both or perhaps something quite different? Does it make sense to separate trees and fields if cultivation is done for a few years only, and would forest vegetation come back in order to be cleared again for subsequent cropping? Are forests and fields not part of the same local space which is used at different times in different ways? After some reflection they may tell us it is more important to distinguish the space formed by the village and its adjacent permanently cultivated land from the surroundings that serve different needs and have a distinct spiritual significance. And they may point out that forests as local surroundings may only be used in accordance with certain customs and with respect and consideration.

Trees and forests supply a large range of products used for food, medicines and daily needs, fodder and local handicrafts, and they contribute to maintain soil fertility for an increased agricultural output. Wood is an important, but by no means the only component providing energy and material for local construction. The variety and importance of uses for local communities and in particular for the poor and disfavoured groups have long been underestimated (FAO 1989c). By referring to forest products other than wood as secondary forest produce, they have been systematically undervalued and neglected. Studies undertaken at the local or regional level show that the economic weight of the non-wood forest products is sizeable and in some cases may be more important than the harvest of wood. The collection and processing of non-wood products provide considerable employment opportunities.

A widespread form of using forests is collective tenure of common property resources (McCay and Acheson 1987). It occurs in agricultural and pastoral societies in which forest and trees are important for the community to subsist and survive. Collective forest tenure does not mean an indiscriminate access to resources. It is based on elaborated tenurial arrangements carefully regulated by customs and rules. It provides different benefits to different groups of people, but even if restrictions for some groups are severe, they do not exclude anyone completely from forest use. This is important in times of scarcity and natural disasters when trees and forests are one of the last resorts for subsistence to peasants and herders.

Collective tenure systems of common property resources have drawn increasing attention from both an economic and an institutional and social point of view. Within the political and economic context which they have developed, they offer a reasonable arrangement for resource management. A combination of restrictions and obligations protects trees and forests to varying degrees and allows for some form of continuing use. However, collective tenure is slow in adapting to change and may lose its regulatory force and provoke exploitative uses if confronted with new forms of resource appropriation (Bromley and Cernea 1989). The appropriation of land through a process of acquiring private ownership rights as part of agricultural expansion, and the claim of forests as a domain of the state are major reasons for the discrimination of local uses, for the disappearance of collective tenure and for the clearing of forests.

In contemporary societies of a Western culture, there may be different opinions about what forests represent, but most people agree on the difference between forested lands and lands without forests. Forests are perceived as a distinct element of the landscape with clearly demarcated borders and a defined range of uses. Their protection and the regulation of access to them are largely determined by considerations of economic utility. Forests are means of production for goods and services satisfying individual and collective needs, and their management can be optimised in relation to prevailing opportunities and requirements. Forests and forest lands are a property which the owners may, within certain limits, use freely. They may be cleared due to collective and individual necessities, and land may be planted with trees if this promises to be a useful undertaking.

The contrast between forests as a means of production and as a particularly valued element of the environment has developed in societies in which culture and nature are often believed to be in opposition to each other. The specificity and the historic dimensions of this thinking become visible, if one compares it with the meaning of forests and nature in societies with different cultures and spirituality (Schefold 1990). In some of them, nature and forests are not considered primarily as resources that may be disposed of for social and economic purposes and can be used at the discretion of individual and collective decisions. They represent a reality with gods and spiritual powers, in which man has to find his place and which may be harmful and at the same time helpful to him. In the thinking of people in cultures with traditional land-use systems nature and man may form parts of distinct realities which leave space for the beings of the present generations and for the world of the ancestors. Forests may represent the space of wilderness which is the realm of the ancestors and which gives cultural orientation and identity to living beings.

Changes of the forest vegetation result from continuing practices of local uses. It requires detailed studies in order to assess their impact on the supposed natural vegetation. More visible and sometimes profound changes are due to intensive silvicultural practices, change of tree species, tree breeding and to large-scale afforestation. If such developments have in the past mainly occurred in Europe and North America they become increasingly important in the tropical and subtropical belt and in particular in Asia and Latin America. Changes in land-use result from clearing forests for agriculture and settlements, grazing and pasture, and man-made forest fires. Large-scale clearing for agriculture and pasture occurred during certain periods in Europe and North America and occur at present in tropical regions with deforestation rates of one to three per cent in some countries. Grazing and forest fires are usually slower and less spectacular processes with nevertheless important effects in the arid and semi-arid zones as well as in mountainous regions.

Local communities with an almost self-sustained agricultural structure and a by and large still sustainable management of the immediate surroundings have a great environmental and political value. This sustainability is not limited to a reflected economic calculation of resource use only, but involves a rather broad and highly flexible mode of subsistence and local self-determination. Joint land management systems and a sustainable use of common property resources require specific frame-conditions in order to be workable, socially acceptable and economically efficient (Ostrom 1995). The definition of the resource has to be unequivocal and the entitled users must be clearly defined; participation of the users in decision-making on common management rules has to be institutionalised and accessible to all; the joint management rules must be subject to an effective control and to sanctions if not respected; there are established conflict resolution processes that have to be used in case of serious disagreements; and administrative action from outside should not obstruct the self-regulatory capacity of the local community managing the resource, or make it obsolete.

Forests as resources for industrial wood production

During the last century a change in emphasis has taken place in the Western hemisphere from multiple forest uses in a local area to a single and economically important use, usually timber production. In the first stage forests were exploited with little regard for their long-term timber production potential and for soil conservation and water regimes. But soon it became evident that the prospects of an expanding wood products industry required a lasting forest productivity, and this needed investment in forestry and an adjustment of timber harvest scheduling.

In Europe this evolution was preceded by a long period in the seventeenth and eighteenth centuries when the sovereign and the nobility had claimed wood resources for operating local industries and long distance trade. They had established juridical control over certain forest lands, forest administrations controlling local uses, and supervision of management practices in communal and, to a lesser degree, private forests. The growing influence of the state on production objectives and management practices led to controversies and resistance in particular on the side of large forest owners. It also created continuous tensions between the forest administration and the peasants and villagers who were more concerned with local uses than with commercial wood production.

The emphasis on the importance of forests for commercial production has had important consequences with regard to the role of communities in the management of resources. In several countries it has been a deliberate and forcefully implemented policy to restrict existing collective uses. Usufructuary rights were abolished in areas classified as state forest domain and collective tenurial arrangements were transformed into clearly defined categories of forest ownership. In some cases this has favoured the constitution of communal forest ownership, whereas in others it has increased the state forest domain. Part of the forests used in common has been distributed among the previous collective users and has come under private ownership. Very often a combination of tenures has developed. The present ownership pattern in European forests is to a large extent the result of this trans-formation process.

Market economies, the change from subsistence to commercial and highly productive agriculture and the introduction of industrial production, have changed the social relevance of forests. They were now considered as a 'resource' with a specific meaning, a resource for the production of wood. Growing wood products markets, based on long distance exchanges, made wood a major national economic option. The concept of forestry as a sustainable timber production activity became the principal objective of forest conservation. Forest policies as a national issue developed in Europe in this context. They were induced by an experience of timber shortages or an anticipated scarcity of future supply in economies that were still depending in many ways on locally available raw material. At the same time the improving economic prospects, demonstrated by new wood products, demands and technologies, as well as by rising timber prices, made it attractive to governments and land owners to promote timber production in expanding market economies. Policy programmes, determined by the legislatures and governments, set the framework for the development of forestry and the wood products industry. The principal actors in elaborating and implementing the programmes were governmental forest services and major user groups such as forest owners and wood products industry associations.

Policy objectives focused on the protection of forest lands, the regulation of harvesting practices, the establishment of a viable sector economy organised by forest enterprises, and the promotion of large-scale afforestation programmes. Yield regulation, silvicultural measures such as thinning and tree improvement, planting of high-yielding tree species and site melioration became a progressive and acknowledged means to increase wood production. To some extent policy targets also included protection, in particular in mountainous areas. National programmes were on the whole fairly simple, but for that reason effective. They prohibited forest clearings, restricted clear cutting, provided technical advice and, to a lesser degree, financial support to forest owners.

A somewhat comparable evolution took place in North America, perhaps with a stronger emphasis on large-scale industrial development. Forest resources were to finance new infrastructures and settlements. Capital intensive, high-yield forestry regimes developed in some regions. Certain forest areas were excluded from commercial wood production mainly as parks. By 1900 the United States had adopted conservation programmes for federal forest lands that combined sustainable timber production with the preservation of certain areas as parks and wilderness. The policy framework that had developed around the turn of the century and continued to evolve, emphasised protection of forests from wildfire and management based on scientific principles. Its implementation implied a strong public sector role in forestry research; the protection of forests from wildfire and epidemics regardless of ownership; the productive management of federal and state forests; the promotion of improved resource utilisation on private forest lands through technical and financial assistance and tax incentives; the adoption and enforcement of state and federal wildlife conservation laws; and the acquisition of public lands for stream-flow protection (Schmithüsen and Siegel, eds. 1997). Policy implementation was largely based on co-operative efforts among federal, state and private sector interests.

In Canada forest policy development by the federal government and the provinces has followed, probably with some delay, a pattern similar to that in the USA. Due to the large proportion of public forest land owned by the provinces, policy and resource management are to a large extent concerned with joint land utilisation and management, involving the public sector as land owner and the private sector in timber production and wood processing (Haley and Luckert 1990). The provinces have developed a range of permits, licenses and forest leases and modified the applicable regulations and contractual arrangements in order to improve forestry practices and to enforce nature and environmental protection.

Under the conditions of tropical and subtropical forest zones, only a small proportion of the exploited natural forests has been utilised in a sustainable manner so far (FAO 1989a, 1989b, 1992). Silvicultural efforts have largely concentrated on plantation forestry which is replacing the natural forest stands in some areas. The priority given to industrial forest development has in many cases detracted governmental action and policy options from local uses that are important to people. This widened the gap between the reality of rural populations strongly depending on trees and forests and national objectives, which too often have only induced an exploitative extraction process. Rotational and combined land uses which are still dominant in many regions of the world have been officially acknowledged only recently and 'reinvented' under such terms as agroforestry, rural forestry or forestry for community development. But such concepts still determine only very few policies at national and regional levels. It will demand more consistent efforts from governments to design and implement forest and rural development programmes which contribute to an improvement in the living conditions of local communities.

A process of state appropriation extending to ownership, usage rights and management control has taken place in many countries. Ownership of forest lands for which no land title could be produced, was claimed to belong formally to the state. The application of this principle, already introduced by the colonial powers and subsequently retained by many forest and land tenure legislation of the independent states, created a new situation (Chakravarty Kaul 1996, Pathak 1994). The governments saw an opportunity to obtain fiscal revenues from commercial logging operations, to promote log and wood product exports in order to improve trade balances, and to develop a modern forestry and forest industry sector. Since forest administrations were usually not in a position to organise large-scale forest resource utilisation, governments relied on the private, and in many cases, the expatriate sector. They usually granted different types of permits and contracts for timber extraction and forest management.

In countries in which the principle of collective local ownership on forest lands has been retained or in which the forest area has been recognised to be under communal tenure, state intervention and more subtle forms of appropriation operate. This refers, for instance, to regulations which invest the right to contract with industrial operators in the state acting in trust for local owners and to sharing of timber revenues between the communities and the government. It also refers to strict management rules which usually limit the decisional powers of the land owners.

The concentration of benefits from forestry at the national level and the centralisation of land-use decisions have important consequences. The groups using forest surroundings traditionally tend to feel deprived of what they consider as their own property. A process of the transfer of resources from the poor to the rich and from rural to urban areas is initiated. The breakdown of customary rules protecting forests as part of the local space and the failure of modern forest laws that were adopted to replace them, may be explained in this manner. The lack of local opportunities to use forests, combined with often short-term industrial exploitation, contributed considerably to deforestation, since this appears to many people as the only possible alternative to draw some benefits from the available lands. If general developments have led from local to national use regulations, many considerations appear now in favour of reconsidering the need for local participation in political decisions and for joint management systems.

The almost exclusive focus on national policy development, regulating forest protection and utilisation, does not correspond any more to the reality of today. Social and political processes at the level of local communities reflecting different interests in forests require more attention and research analysis. National programmes can only be successfully implemented, if they are meaningful to, and accepted by, local groups and communities. At present, forests have become a world-wide public concern and subject to political efforts in order to develop consistent Cupertino for their protection.

The perception of forests as environment and wilderness

During the last three decennia, a wider perspective on the value of forests has emerged with important challenges to forestry and forest policy. The first momentum came from the rapidly increasing demands for recreational activities in open landscapes and forests (Schmithüsen *et al.* 1997). A second momentum, which now has a world-wide dimension, developed from the environmental and nature-protection movement. A third element of change superimposed on the other two derived from the interest of people in forests as an element of personal identification with wilderness as imagination of a cultural value and a spirituality of freedom and contemplation.

In 1992 the United Nations Conference on Environment and Development put forests and forestry on the international political agenda. The preparatory stages as well as the follow-up processes to this conference, represented by the work of the UN Commission on Sustainable Development, show that considerable disagreement on many issues continues to exist, and there is a reluctance, in spite of a growing public awareness, to take forests as a common world heritage. Large-scale deforestation in the tropics and sub-tropics lead to social and economic problems for many people living in these regions. Protection of forests, sustainability of uses and preservation of biodiversity are challenges of equal importance in many industrialised countries in the boreal and temperate zones. The conflicts which arise are part of international controversies concerning social justice, self-determination and sovereignty, local political participation and democratic decision-making, as well as symptoms of the unbalanced economic development between different parts of the world.

New elements that appear are an evolving perception of citizens on the urgency of conservation needs and to maintain future options. There is an increasing involvement of interest groups articulating conservation demands more forcefully, and growing insistence among the public to see more effective policy measures for the preservation of nature and landscapes. The environmental movement has also made it clear that espousing conservation as a principle does not necessarily prevent degradation and waste of resources. It calls for specific measures and regulations based on a more precise understanding of how resources can be used in a sustainable manner and which elements of our environment must be protected and preserved. Value changes from a largely resources utilisation oriented perspective to an environmental perspective influence future management options on forest lands increasingly. This process leads to policy changes and new legislation as well as to an increasing influence of local groups on forest management decisions.

In a global perspective forests and woodlands are perceived as an important element of the natural and social space in which societies develop. Their potentialities leave options to present and future generations, some of which we may already be aware of and some of which we cannot yet foresee. In such a perspective trees and forests are a representation of nature which is supposed to be largely free from apparent human disturbance, some free of it altogether. Forests mean unfettered natural processes and wilderness, something very different from surrounding intensive urban and rural land-use patterns. The efforts to limit forest uses, to set aside areas without apparent human disturbance, and to preserve biodiversity in managed forests are the result of a growing concern to maintain forests as a representation of nature and wilderness.

These developments, representing a fundamental current in many societies, have put forests in a much broader political context. The evolution of a network of policy programmes addressing conservation and management corresponds to evolving perceptions and new political demands. In addition to sustainable wood production, forest policy considers the objectives of infrastructural protection, recreational use, nature and landscape protection and spiritual and aesthetic values more systematically. These objectives are increasingly superimposed by those of cross-sectoral policies, addressing nature and landscape protection, environmental protection and land-use planning. In the beginning, they concentrated on general principles of resources utilisation, but over time they also became concerned with the regulation of specific forest uses. As a result, forestry matters are now the subject of several programmes which for historical reasons have developed in a different manner and which, to some extent, compete with each other. Sectoral forest policies tend to emphasise production and economic land-use aspects, while cross-sectoral policy objectives provide benchmarks for forest protection and environmental conservation.

Strategic value of socio-cultural research on local knowledge related to sustainable forestry practices

A wealth of knowledge, transmitted from generation to generation, has been accumulated indicating the meaning of trees and forests and the uses of particular species and products as well as the technologies which enable such uses. Experience has taught users how trees and forests should be managed in the interest of the community and how to develop practices that ensure their protection. Research on local knowledge of conservation practices shows its ingenuity, cultural distinction and social usefulness. Local, or in a more specific sense, indigenous knowledge is of particular importance in traditional societies because the local context is an universal frame in which knowledge matters.

Research on indigenous knowledge and on the meaning of trees and forests in different cultures matters and is necessary in order to show the many needs and values associated with forests in a given locality, and the importance of forested areas for the survival of indigenous people and their cultures (Seeland (ed.) 1997). Research findings bring forward concrete facts to be considered, which result in political decisions being made on the utilisation of natural resources, and they are also a strong argument for increased efforts in protecting forested areas and considering local uses. These findings also demonstrate the confrontation between customary performances in sustainable resource management, developed from experiences and adjusted to difficult living conditions, and the development activities of modern states (Bharara 1999).

In the interface of tribal and non-tribal sections of the society it becomes more and more important to outline and stress the wealth of knowledge and the merits as well as the problems that have to be taken into consideration on behalf of local forest dwelling communities. Most important is a reassessment of the possibilities of integrating the forest related knowledge of the tribal population into land use and management programmes. Through research it gains a lot of prestige and political empowerment, and raises the self-esteem of a notoriously underrated section of modern society. This is a major aspect of research on indigenous knowledge. Another aspect is the entitlement to participate in decision-making processes at various levels with the help and mediation of local and regional NGOs. This goal can only be achieved if the political bargaining power of tribal populations is upgraded through research on the potential of their knowledge so that they may become a more acknowledged partner in environmental planning processes.

There exists an increasing awareness that research of local knowledge of forests is relevant in all countries. The advancement of knowledge, induced by changes in attitudes and perceptions, may help to adjust the prevailing use patterns in a social evolution of changing potentials of resources (Burch and De Luca 1984).

In many cases local knowledge and experiences are not considered, but instead differ from national resource concepts and governmental management regulations. The results are obvious: local knowledge and social obligations are bypassed, and the management schemes introduced from outside remain inapplicable and of little significance in the local setting. Investigations into the perceptions and attitudes of people and local communities with regard to the conservation of trees and forests have therefore an increasing bearing on forestry development and policy. They provide the possibility for an improved understanding of existing management practices under their specific social and cultural conditions, as well as the potential for adaptation to a changing economic environment.

Research on the prevailing socio-cultural values and uses, as reflected by local knowledge, is important for the efficiency of participatory resource management policies (Poffenberger and McGean, eds. 1996). In increasingly urban and peri-urban societies, knowledge and personal experiences of forests as influenced by evolving attitudes and perceptions are particularly in a state of change. Available research results indicate that knowledge on traditional uses is still present, but fading and with little relevance to the context in which actual problems of protection and management are seen and assessed. Educational activities in forests try to introduce people to the abundance of nature and to make their surroundings and history more meaningful to them. There is also a need to increase the understanding of forests as a source of raw material and of forestry as a productive land-use activity. This aspect cannot be dealt with only as expert issues left to the small group of specialists competent in forest management. The indications from current studies on public attitudes towards forestry confirm the need to place timber production more clearly in the context of global sustainability and to explain more precisely its social usefulness in utilising renewable resources.

The process of adapting forestry practices to a broader range of social demands and of developing management systems, adjusted to varying local conditions, does not take place without difficulties and results in divergent positions. Multidisciplinary research has to embrace the cultural and social factors that influence individual and collective forest uses, attitudes and perceptions of people on various options in forestry development and the political processes that occur at all levels of the community (CIFOR 1995). The meaning of forests in different societies and at different times and the performance of a society with regard to use and conservation patterns become central issues of an innovative socio-cultural research activity. This implies an extension of the research agenda in order to gain a more fundamental understanding of the political relevance of forests as related to specific socio-cultural values. A broader context is indispensable in order to compare different situations, to discuss measures for the conservation of forests, and to evaluate experiences in forest policy developments. This kind of research examines the changing role of forests for their respective societies, the variations of the political context and the common denominators which determine conditions for their conservation.

The intercultural dimensions, in which trees and forests are important, require that professional attitudes comprehend social and cultural information in order to become appropriate contributions to sustainable management of renewable natural resources. Research has to put more emphasis on the manifold ways man performs in his environment. This calls for new forms of collaboration between the social and cultural sciences and established networks in applied natural sciences and forestry research. It implies a more intensive exchange of ideas between researchers who emphasise the social dimensions of forestry, and those who are *ex officio* responsible for resource protection and development.

Conclusion

The importance of a forest is reflected in its cultural perception, and societies have developed their notions of what forests mean to them. In order to know something about them, the political context of these cultural perceptions has to be investigated, particularly in times of the rising importance of sustainable forest management at the international level. At local levels multi-disciplinary and cross-cultural research is necessary to obtain detailed data on the local people's use and meaning of their forest environments.

Research programmes on forests and trees that make a useful contribution in a changing social reality have to embrace the full perspective of forests as an essential, many-faceted component of man's environment, as well as of forestry as an economic and production-oriented mode of land-use. It is this two-fold relationship that research has to consider by analysing individual and collective forest uses, attitudes and perceptions with regard to various resource options and contractual forest management arrangements involving different sections of the community. Inquiries into the many forms of local knowledge in traditional and modern societies are an important subject in socio-cultural research on trees and forests. Its results are a strategic element in promoting sustainable forest development based on participatory utilisation practices and joint natural resources management systems.

References

Bharara, L.P., 1999: *Man in the Desert - Drought, Desertification and Indigenous Knowledge for Sustainable Development*, Scientific Publishers, Jodhpur, India. 394 p.

Bromley, D.W.; Cernea, M.M., 1989: The Management of Common Property Resources, World Bank Discussion Paper Nr.57, Washington D.C.

Burch, W.R.; De Luca, D.R., 1984: *Measuring the Social Impact of Natural Resource Policies*, Albuquerque.

Chakravarty Kaul, M., 1996: Forest Rights and Forest Laws in the Indian Himalayas during the Second Half of the 19th Century, *Forstwissenschaftliche Beiträge der Professur Forstpolitik und Forstökonomie der ETH Zürich*, Vol.16: pp. 42-63.

CIFOR, 1995: *A Vision for Forest Science in the Twenty-First Century*, Center for International Forestry Research, Bogor, Indonesia, p. 40.

FAO, 1989a: Review of Forest Management Systems of Tropical Asia, FAO Forestry Paper 89, Rome.

FAO, 1989b: Management of Tropical Moist Forests in Africa, FAO Forestry Paper 88, Rome.

FAO, 1989c: Forestry and Food Security, FAO Forestry Paper 90, Rome.

FAO, 1992: Management and Conservation of Closed Forests in Tropical America, FAO Forestry Paper 101, Rome.

FAO, 1994: Readings in Sustainable Forest Management, FAO Forestry Paper 122, Rome. p.266.

Firey, W., 1960: *Man, mind and land. A theory of resource use*, Free Press of Glencoe, Illinois.

Gregersen, H.; Draper, S.; Elz, D., (eds.), 1989: *People and Trees. The Role of Social Forestry in Sustainable Development*, Washington D.C.: World Bank.

Haley, D.; Luckert, M.K., 1990: Forest Tenures in Canada — A Framework for Policy Analysis, *Forestry* Canada.

Harrison, R.P., 1992: Forests, *The Shadow of Civilisation*, The University of Chicago Press, Chicago and London.

McCay, B.; Acheson, J. (eds.), 1987: *The Question of the Commons*, University of Arizona Press, Tucson.

Ostrom, E., 1995: *Governing the Commons. The Evolution of Institutions for Collective Action*, Cambridge University Press, Cambridge.

Pathak, A., *Contested Domains — The State, Peasants and Forests in Contemporary India*, Sage Publications, New Delhi, p.172.

Poffenberger, M.; McGean, B. (eds.), 1996: *Village Voices, Forest Choices — Joint Forest Management in India*, Oxford University Press, Delhi/Bombay/Calcutta/Madras, p. 356.

Schefold, R., 1990: Natur und Kultur im Weltbild indonesischer Stammesgesellschaften. In: Boehm, W.; Lindauer, M., (Hrsg.), 1990: Woher, Wozu, Wohin? Fragen nach dem menschlichen Leben, S. 285-308. Klett, Stuttgart.

Schmithüsen, F., 1996: The Meaning of Forests in a Perspective of Social and Political Development. In: L'uomo e la Foresta — Secc. XIII-XVIII. Serie II — Atti della "Venrisettesima Settimana di Studi", Instituto Internazionale di Storia Economica Francesco Datini, Prato, pp. 79-97. Le Monnier, Firenze.

Schmithüsen, F.; Siegel, W. (eds.), 1997: Developments in Forest and Environmental Law influencing Natural Resource Management and Forestry Practices in the United States of America and Canada, Selected Contributions submitted to the IUFRO Research Group 6.13.00 Forest Law and Environmental Legislation, IUFRO World Series, Vol. 7., Vienna, p. 373.

Schmithüsen, F.; Kazemi, Y.; Seeland, K., 1997: Perceptions and Attitudes of the Population towards Forests and their Social Benefits: Social Origins and Research Topics of Studies conducted in Germany, Austria and Switzerland between 1960 and 1995, International Union of Forestry Research Organizations IUFRO, Occasional Paper 7. pp. 1-30, Vienna.

Seeland, K. (ed.), 1997: *Nature is Culture. Indigenous Knowledge and Socio-cultural Aspects of Trees and Forests in Non-European Cultures*, Intermediate Technology Publications, London, p. 152.

3

What is Indigenous Knowledge and Why Does it Matter Today?

Klaus Seeland

General considerations

Two decades ago (Chambers 1979) indigenous technical knowledge became an important topic in the debate on an economic and culturally sound development of Third World countries. Ever since then it has been discussed predominantly with regard to developing appropriate technologies instead of transfer of modern technologies and ways of applying them in the agrarian sector by reintegration or reinvention of what once has been part and parcel of a self-sustaining mode of production. The most prominent example that resulted from this discussion is agroforestry, where traditional agriculture, forestry and livestock-management make an integrate model of primary production under conditions of traditional societies. Agroforestry is an example of a new term for a more or less age-old practice to harmonise and maximise the degree in which different sectors of rural economies are integrated in terms of productivity and elasticity of peasant households. As agriculture was and still is considered to be the main source of subsistence in traditional and transitional societies with a large proportion of rural population, it was expected from this discussion that it would provide solutions for the deficiencies that arose with the introduction of the modern know-how of those favouring agrobusiness. The advantages of the Green Revolution, the mechanisation of agriculture and the rising importance of a market economy were to feed the ever rising number of the poor and low-productive rural population.

Indigenous technical knowledge in these discussions did not refer in the beginning, as far as I can see, to forests or forest management. Now that the need for a sustainable management and conservation of the world's forests has gained more and more attention, the interest to use local indigenous knowledge to achieve this goal has increased. The term 'technical' has been skipped in this context. This was perhaps due to the realisation of the fact that it is hard, if ever possible, to use indigenous knowledge in agriculture, as it was hoped, in the purely technical sense the term evokes in a Westerner's mind.

What is indigenous knowledge?

Indigenous means that something is originating locally and performed by a community or society in a specific place. It emerges as peoples' perceptions and experience in an environment at a given time in a continuous process of observation and interpretation in relation to the locally acknowledged everyday rationality and transcendental powers. The context of local social performance makes sense between people who share a common rural habitat, language and knowledge, be it exoteric (open for all) or esoteric (secret knowledge). Indigenous knowledge is human life-experience in this distinct natural and social compound, within this unique local and contemporary setting. In a traditional society the local context is taken as the universal frame in which knowledge matters. This context is constituted out of physical facts, social interactions among people in the surroundings they perceive as their world and spiritual beliefs. Indigenous knowledge is not formally taught, but perceived in a particular context at a certain stage of the perceiver's consciousness that grows in the world of local events.

Knowledge is to be called indigenous, if it is bound to local experiences and takes its local world perhaps not as the only one existing, but as the most relevant of all. It is an authentic appropriation of being. The environment is not an environment in a technical sense, but a world. The outlook that there is a locally meaningful world characterises an indigenous world-view. Local knowledge, as we may also call it, is an encompassing whole of what has been revealed to human perception in a particular place or region. There will always be local knowledge even if modern scientific knowledge invades a socio-cultural setting and challenges its local knowledge more or less effectively. Global knowledge, once it has settled down in the locals' minds and is disconnected from the international setting it refers to, becomes a local variety of what was once global knowledge. A local setting transforms outside knowledge in a continuous process of adaptation and appropriation.

The appeal of indigenous knowledge

It is a matter of fact that indigenous knowledge is an amalgamation of what is perceived as 'here-and-now' important by the local population. The term itself is an ascription from outside, coined out of the perspective that there is paramount and presumably supreme general, i.e., modern knowledge. This modern, or better to say, non-traditional knowledge is distributed by formal standardised Western patterns of education all over the world. However, indigenous knowledge is not an ubiquitous know-how everybody has and neither is it shared equally by all inhabitants of a locality nor is it a standardised and comprehensive account of 'what is known' in a particular socio-cultural setting. Consequently, nobody knows what it exactly is, but a rising number of scientists and policy-makers are convinced that it is an important component to safeguard the resources of a region by managing them adequately in the local manner — the way it has always been done. Whatever it may be, important are the ideas leading to this apprehension.

Firstly, there is the idea that an environment can only be properly managed by its inhabitants and not by ultimately absent experts from outside. The go-on-travelling experts may know 'solutions' to environmental or social problems because they are not part and parcel of the local setting. They do not live where they work for long or even for generations in a tight social cohesion or mutual dependence. The dynamics of social reciprocity in a poor and marginal community is almost beyond what an upper middle class expert from far away places can imagine. But these dynamics shape reality and not environmental rationalisations. Provided these inhabitants are in a position to exercise an active and almost independent political control and regulation over their territories and resources, their indigenous and sometimes age-old way of controlling the use and abuse of resources would be considered appropriate. They are also assumed to be sustainable, as long as an equilibrium between their constituting and stabilising factors prevails.

Secondly, the technical experts and extension agents, who are trained in a Western scientific rationality, are more and more convinced that indigenous knowledge fits very well into a participating approach that involves people with their knowledge in the process of their economic development. Be it as it may, traditional knowledge is in the process of entering into modern development strategies through the back-door. In the light of interest that experts take in indigenous knowledge on a broad scale, it becomes paradigmatic for a new development idea. Why do internationally active elites invest intellectual energy and financial resources into this sphere of interest? Why do, for instance, agencies with a predominantly technical profile like GTZ become interested in indigenous knowledge? The answer may be that, after decades of extension activities, now there has evolved an awareness that modern global expertise has always to cope with autochthonous knowledge that is deeply woven into the social fabric of a local and particularly rural community within a transitional developing society. There is always so to speak a basic layer of knowledge in every community from times immemorial that is a representation of a society in its typical natural surroundings. The wisdom and experience of generations reflect the way of life of their ancestors before one's own generation and oneself took over. Indigenous knowledge is a matter of continuity, although one's surroundings may change at a slower or quicker pace. It is a matter of social and cultural evolution in the background of an ancestral tradition.

Sooner or later, tribal communities respond to outside or social mainstream developments in either a conservationist, adaptationist or fundamentalist way, that in many cases may lead to political antagonism with the dominant society. This response reflects the respective status and political power of a community or society and its inner strength, self-esteem and capacity to eke out a living even under circumstances of environmental stress and ethnic harassment. One question of strategic importance therefore is, whether or not the propagation of indigenous knowledge might be an obstacle to social evolution itself? An obstacle in the sense of attracting the attention to knowledge that has been important for the past but may now distract the view from the demands of the future.

To develop its own capacity to cope with the challenges of environmental stress and social marginalisation may be a community's prominent goal. The achievement of this goal decides whether its socio-cultural identity would persist or an integration into the social mainstream of a mass society in transition will happen. It remains to be seen whether traditional ways of life and indigenous knowledge can be revitalised or operationalised within the context of development. We do not have enough experience at the moment to predict as to what results international attention, financial support, research and political encouragement will lead to. If after some time these communities and NGOs are left alone with their efforts and enthusiasm to uplift the living conditions of poor marginal communities with the help of their own traditional concepts are it will be seen how sustainable they are.

That these concepts have largely failed so far in a contest with so-called modern knowledge is more or less evident, for instance, for at least those poor members of a forest-dwelling community who have seen a bit of the wealth of the cities. Why should they cling to their traditional knowledge and lifestyles? Why should they keep a way of life that has proved to be inferior in achieving technical assets and all the amenities of the agrarian or urban middle classes? Why do some of the Western technical experts favour a knowledge that stands for a social status that is widely considered to be economically underdeveloped?

The appeal of indigenous knowledge lies for once in its underlying assumption that a bottom-up strategy including locally available capacities concerning the management of resources is always better than to transform local villagers or forest nomads into clients of the Forestry Department or any Government service. The transfer of purely technical know-how is considered not to lead to a sustainable use of resources.

Forests, trees and their products are to be managed by indigenous people with the help of their knowledge for local consumption in largely self-sustaining communities. Scientific forestry with its specialised knowledge manages forest resources, predominantly timber for the market. The dilemma of the two ways to deal with forests and their products is that one type of knowledge serves one purpose only. Local self-determination in the supply of locally available resources is contrasted by the needs of the ever-rising population in the mass-agglomerations of the cities that have to be met with products from the forests as well. Consequently there is an idea of decentralisation and local autonomy to regulate the use of any resource behind the argument to favour local forms of forest management for the local population instead of serving the economical purposes of city-dwellers and wood-traders. However, what seems to be the most relevant appeal of indigenous knowledge that emerges out of the principles of a local subsistence economy is independence. The use of local indigenous knowledge does not turn independent forest dwellers into clients of the state administration, at least not the elder ones, because they stick — right or wrong — to their own traditions and customs. In a phase of de-regulation and decreasing administrative capacities an independent and self-sustaining social performance matters very much. Low-cost-considerations and the recognition of the important principle of subsidiarity have its appeal for policy-makers and development experts.

In a society in transition, market economy and a remarkable division of labour has entered social life even in the remotest pockets of the country. And division of labour means as well division and exchange of resource supply from other than local communities mostly represented by the state and commercialists in the first instance and later on followed by other sections of the society. In a political perspective there has to be a balance of commercial exploitation and an environmental as well as a social consideration of the use of resources for mass consumption. The ethnic and natural diversity of a subcontinent such as India shows, for instance, the different interests of people in remote and sparsely populated regions with dense forests and overpopulated areas where there is practically no more tree cover at all. The inter-connectedness of the interests and regions has to be seen in a perspective of balancing the environmental weights of development politically. Local and global knowledge ultimately have to share in complementary roles to serve various interests of resource utilisation, production and consumption within the larger setting of the multicultural Indian society.

Important questions concerning local indigenous knowledge

Resources may be defined as local wealth that is, however, not always available to the local inhabitants. Their incorporation into a state society and the respective state authority that is concerned with the administration of resources decides upon its use, exploitation and management. This relates first of all to those resources which are underneath the earth or technically difficult or costly to exploit. Indigenous knowledge is a sort of 'soft exploitation' of what is at hand to bypassing nomads or hunters and gatherers who have acquired an intricate knowledge about the use and management of several items over generations. They make use of what they take for granted to them, i.e. what is useful and manageable for them at their capacity level. The knowledge applied in this process reflects the state of a community's rationality and its application at the lowest available social cost. It also sheds light on religious restrictions and taboos that are connected with particular aspects of an environment. Thus the question for a tribal forest dweller is not 'to use or not to use', but to use in social and spiritual accord with the prescriptions of his culture. The conduct of his lifestyle is thus a representation and pattern of resource use, which he would rarely alter or transgress without being forced to do so.

For development approaches trying to integrate these patterns of traditional resource use into a multicultural concept of civilisation the question arises, *whether indigenous knowledge is to be equated with ecologically sound knowledge*? There is, for e.g., an overall notion of cosmos in any tribal society that encompasses all physical, social and spiritual aspects of life in a form of animism or natural symbolism and ritualism. This does not mean that all interactions between physical aspects are known to rural people as interconnected in the sense of the term in natural science, 'ecological'. If one, however, understands ecology as "to know about the interconnectedness between all forms of life as such", then indigenous knowledge can be called ecological. Yet ecological does not necessarily mean that tribal forest-dwellers care for a sustainable resource use — that is, sustainability as emphasised in the recent Western environmental discussions. Ecology should much more be understood as the reproduction of a cultural lifestyle that has been passed on and not as an economical rationale. It is an open question whether the alteration from outside that is forced upon a forest dweller's community is a much more serious threat to it than the adaptation of this culture to the changing environmental conditions. On the whole, I might say, a member of a tribal community living in the forests is very well aware of natural processes that happen in the jungle, but cannot (or dares not) alter these processes because he considers himself not as an authorised change agent but as a part of his cosmos, who himself is directed by spiritual and cultural powers in and around him. Consequently technical experts should not trap themselves in an imagination of 'the good and ecological wise savage'.

Researchers and development executives are both keen to know, whether indigenous knowledge, as far as its environmental aspects are concerned, *can be transferred out of its local context into another one and used, for e.g., for a regional inter-ethnic, national, or even international network*? From what has been said above, it seems obvious that indigenous knowledge cannot be displaced, because it is autochthonous empirical experience and only valid as a *world* in its place of origin. Authenticity means to know things from personal experience in places where they originate and thus experiencing them in their proper context. What can be transferred is the *rationale*, how indigenous people deal with nature and cope with times of scarcity of resources and social competition to gain access to other means of livelihood. What else is comparable is the way of looking at natural, social and spiritual phenomenon in an ultimately inseparable way. The modality, how tribal forest dwellers perceive things, live with them and use them is a mode of appropriation that they may share with many other tribals all over the world. To collect them and use them for a network presumes that rural dwellers can anticipate what may be useful for them. It contradicts the principle that those who know and what is of local relevance to be known is part of the same environment. Hence as man's thinking is inspired by the whole physical, social and spiritual world he is living in, it would be impossible for him to put in a nutshell what has been appropriated individually over one's lifetime and make it a transferable 'information'. What occurs to a person in his daily life or is passed on as knowledge in his culture from times immemorial cannot be easily applied in another context. His knowledge is appropriated authentically, but in other places his knowledge may not be effectively applied, because the circumstances and conditions of these places are different from those where his knowledge has been generated.

Another question that is discussed vehemently is, *whether indigenous knowledge is common property, and whether individuals – either tribals or non-tribals – can claim intellectual property rights over it*? Knowledge is common as far as it deals with common everyday life that everybody knows. But it should not be called a common property, although an insider can hardly be excluded from this knowledge. Experts, say anthropologists, who know some of a tribal's standard knowledge, can appropriate it individually, write books on it and sell the information to their readers. As authors they can claim copy-rights. Nobody, however, neither an insider nor an expert outsider can claim property rights of what is thought or done in the everyday life of a region. 'Professional' or specialist's knowledge, emphasising special skills, is usually the 'property' of a family, a clan or in a more stratified society, a guild. The appropriation of intellectual property rights is based on proto-economic thinking which claims that ways of thinking or social performance in general are or can be commercial commodities. Indigenous knowledge is a way of social performance of what is at the disposal of individuals in a community under conditions that would represent its typical cultural pattern. Social performance denominates the options of individuals or social institutions to act within the guidelines of cultural "do's" and "don'ts". For a folksong for instance nobody can claim a copyright and ask for royalties, if it is sung anywhere in the world. The same applies to Swiss cheese, English marmalade, French fries or Indian tea, etc. Indigenous knowledge in general is an authentic world heritage of particular cultures, who have contributed to it in their characteristic ways.

If one asks, *whether indigenous knowledge can be taken as an alternative to the technologically dominated mainstream development*, the answer from my point of view should be definitely negative. Indigenous knowledge has a distinguished profile and it cannot be replaced by nor can it replace other forms of knowledge. Different sorts of knowledge may contradict each other particularly in the fields of primary production such as agriculture, fishing, forestry, where various cultural concepts have led to different rationales over time. The mainstream of internationally valid technological knowledge is free from any cultural attachment and is a claim of professionals from whatsoever culture they may come. Agricultural extension in agroforestry for instance is applied by international experts in any country of the world. The actual contradiction between global and local knowledge is that global knowledge draws its legitimisation from its claimed technical effectiveness, which is not linked to any specific place and cultural setting. Local indigenous knowledge, however, is 'work to measurement' and only effective in its cultural context. Global knowledge makes sense if it is most effective to achieve a purpose or a technical objective; the sense of local knowledge lies in its appropriateness concerning the whole cultural context it refers to. If it is made a technical or mechanical 'tool' it may never be effective or sustainable in its achievements. Competition between both forms of knowledge for resource utilisation and sustainable forest management has to be seen in the background of different needs and interests of users. As I argued above, different sections of a society try to fulfil their respective demands and needs by applying different sorts of knowledge. Subsistence farmers or forest dwelling nomads do it naturally with the help of their locally acquired traditional knowledge; town dwellers make use of the highly stratified professional know-how that is provided in densely populated and economically developed areas. If one looks at the situation from the point of view of a resource supply that has to take the needs of a larger state unit into account, the use of knowledge has to be seen complementary, according to the context in which each form of knowledge has been appropriated.

Why does indigenous knowledge matter today?

Indigenous knowledge matters today because it covers a very important aspect of the survival options for subsistence communities in the twenty-first century. In a process of world-wide deregulation the administrational capacities of the state are supposed to be no more sufficient to handle and tackle the major problems of large populations, who have become in one way or another dependent on the services of the state. The problem- solving capacity of the state in transitional societies of Third World countries is decreasing. This is not only due to the notorious financial crisis of the state, but the political legitimacy of democratically elected governments and their administrations is being increasingly challenged and disputed. Communalism and strong ethnic or religious movements propagate a separatist policy in a struggle for political autonomy. In the wake of these political developments the role of forests has become highly political. Forests have only partly been integrated into the socio-economic mainstream, highly productive agricultural, industrial or settlement areas. They have been so far claimed for wood production through scientific forestry over the last one hundred years in a process of state appropriation thus reducing the possibilities of a self-sustained existence for local forest dwellers. In this context it is important to realise that state power still enacts its political mandate over the forests to adjust them and their inhabitants to the rationality of markets and the regulations of administration, although there is a common trend towards deregulation.

The political dimension of forest related indigenous knowledge

Environmental problems do reflect the state of political controversy, in which various sections of a society are involved according to their respective interests and political bargaining power. Usufructuary rights and benefits connected with the use of resources are subjects of such controversies. If they relate to resources to be taken from forests, i.e., from areas of national interest and importance, these controversies indicate a lack in political consensus over the role of the state as a political representative of the distribution of national wealth. This lack of consensus, for instance, between ethnic groups and the social mainstream, is shown by some political groups, who are emphasise conservation and working for the survival of poor and marginal sections of the society and others who favour the economic development interests of the advanced sectors of the society in a nation-wide struggle over the appropriation of resources.

The recent history of the expropriation of tribal societies in post-colonial nations is in most cases a depressing account of an environmental colonialism, where ethnic sections of young developing nations are pauperised, expelled from their territories and vanish with the forests they used to live in. To read the great and still increasing number of books and many documentations of misery and poor living conditions in tribal societies makes the reader uneasy and raises protest. These are people who to a large extent still live a life very much exposed to the powers of ruthless nature and live at the mercy of the hierarchies of administrators in modernising societies who are more interested in the achievements of urban middle class development standards in countries which are inhabited by an overwhelming majority of rural masses.

The knowledge of forest dwellers or poor peasants is the legacy of a cultural heritage and tradition that shaped the face of the forests and from where some of today's cultures took their origin, but is at the crossroads of a new era. Concern for ecology and environment reflects a commitment of a world community that considers itself to be aware of the linkages between knowledge and conservation. Ecology and sustainable management of the environment are but other terms for the common future of all people of this globe favouring diversity and pluralism against 'monoculturalism'. However, national and international politics indicate that the struggle for survival and control over the use of resources inside and outside the boundaries of a country is perennial. The dispute over what is looked at to be a resource for the whole society and what is left at the disposal of those ethnic groups who claim ancestral rights over it, is socially contested. The role of indigenous knowledge is to be seen in the perspective of this contest, because it cannot transcend it. It is up to a society's political will to acknowledge and safeguard the value of the country's spiritual and intellectual heritage not as a goal in itself, but for the sake of having common future options. Forest management as a contribution of tribals to a national environmental policy may be a possibility to share social responsibility for a country's forests and its renewable resources.

Conclusion

The appeal for an integration of indigenous knowledge into the development process of a society in transition would be naïve if it remained blind to the massive political obstacles it is confronted with. Contemporary environmental problems ask for solutions which are a match to them. Although it may sound odd to some people, it is widely accepted among development experts that local knowledge is a valid match for contemporary problems, because it is not merely environmental knowledge in a technical sense. It is accumulated wisdom of cultural traditions that has messages for those who are prepared to listen to them. Although they lack essential political bargaining power, tribal people and their knowledge may survive and be helpful in times of administrational deregulation and a self-responsible management of resources in the regions where they live. One may have doubts as to whether scientific knowledge alone will be in a position to become 'socialised' through formal education one day. There will be no, as I see it, common knowledge of all for all in one world in the twenty-first century to solve, among others, the environmental problems which future generations will have to face. As I put it before: environmental problems have to be tackled locally and by those who live in their dwelling place over a longer period of time. Their commitment to claim their chances of survival with the help of what they know and how they perceive their situation will ultimately count and decide the destiny of people in their respective environments.

This plea for research and the propagation of indigenous knowledge to contribute to a social and culturally more appropriate way of management of local forests is made from a social scientist's position and aims primarily to stimulate local non-governmental organisations (NGO) to integrate people's knowledge as a major component into their work. Indigenous knowledge operationalises itself to the extent tribal people are allowed to live their way of life according to their own ways and values. It matters most, what they consider to be useful for them and what an obstacle to their social development. It is not that tribals should stick to their indigenous knowledge only or be excluded from other sources of knowledge they might consider necessary to cope with modernity which is invading their world. The decision to face the future with a cultural identity that has evolved from a tradition that has developed from having been exposed to the hardships of particular natural surroundings has to come out of a self-esteem and conviction of the appropriateness of one's knowledge to meet the challenges of one's own future.

References

Bahuchet, Serge (coord.) (1993): *Situation des populations indigènes des forêts denses humides*, Paris: CNRS, Laboratoire de Langues et Civilisations à Tradition Orale.

Brush, Stephen B. (1993): Indigenous Knowledge of Biological Resources and Intellectual Property Rights: The Role of Anthropology, *American Anthropologist* 95 (3), pp. 653-86.

Chambers, Robert (ed.) (1979): Rural Development: Whose Knowledge Counts? *IDS Bulletin*, vol. 10 no. 2.

Elwin, Verrier (ed.) (1963): *A New Deal for Tribal India*, New Delhi, Ministry of Home Affairs, chap. 12, Forest Policy and the Tribes.

ICIMOD (ed.) (1994): Indigenous Knowledge Systems and Biodiversity Management. Proceedings of a MacArthur Foundation — ICIMOD Seminar, April 13-15, Kathmandu (Nepal), ICIMOD.

Naess, Arne (1990): *Ecology, Community and Lifestyle: Outline of an Ecosophy* (tr. & revised by D. Rothenberg), Cambridge: Cambridge University Press.

Pathak, Akhileshwar (1994): *Contested Domains. The State, Peasants and Forests in Contemporary India*, New Delhi: Sage Publications.

Thrupp, Lori A. (1987): Building Legitimacy of Indigenous Knowledge: Empowerment for Third World People or "Scientized Packages" to be Sold by Development Agencies? Unpubl. manuscript, Institute of Development Studies — University of Sussex.

Woodley, Ellen (1991): Indigenous Ecological Knowledge Systems and Development, *Agriculture and Human Values*, vol. VIII, nos. 1/2, pp. 173-8.

4

Explorations into the Conservation of Indigenous Knowledge for Resource Management in India

Sonja B. Brodt

Introduction

THIS chapter explores some of the conditions and mechanisms by which indigenous knowledge of crop and tree cultivation is either conserved or eroded in Central India. Many scholars (Gadgil and Guha 1995; Gupta 1996b, 1997; Mahanti 1997; Shiva 1997) have expressed a concern that the knowledge developed within India's diverse communities is threatened by globalising, scientising, and therefore homogenising world trends. Considering the now widespread recognition of the value of indigenously developed knowledge of farming and forestry, it is important to ask how such knowledge is faring as it comes into increasing contact with the global economic and scientific system.

Drawing on field observations carried out in 1995-96 in a small group of rural communities in Madhya Pradesh, this study suggests that the types of knowledge used, retained, and passed on is contingent on many interrelated factors, foremost being communication and learning. These two factors, in turn, are mediated by variables such as caste and class and the different levels of social prestige associated with them. In addition, how knowledge is connected to specific places and used at different spatial scales on the landscape may also influence its durability. Finally, the structure of the knowledge itself and how it is utilised by people, including its degree of generality or specificity and its adaptability to outside influences, can be key factors. This analysis suggests that the decline of indigenous knowledge in resource management, while a valid and urgent concern, is not a certain and uniform process. Rather, diverse influences on knowledge production and dissemination are operating simultaneously and may have conflicting effects on the durability of indigenous knowledge. The conservation or decline of such knowledge may therefore occur unevenly over space and time.

The fieldwork for this study focused primarily on technical and conceptual aspects of trees growing around homes and on crop fields. In addition, some emphasis was placed on the use and management of the surrounding forest areas. Knowledge of crop management was also frequently addressed as an easier means of accessing ideas about plant growth and learning strategies, since crops play a much more prominent role in the day-to-day economic lives of most villagers than do cultivated trees. The many similarities between growing domesticated trees and growing crops also result in many overlaps in the respective bodies of knowledge. This chapter therefore addresses knowledge systems for managing cultivated plants in general, with special emphasis on knowledge of tree cultivation.

Social and physical setting

The villages under study are located in the Chiklod area of Raisen district, approximately 40 km south-west of Bhopal, the capital city of Madhya Pradesh, and they are connected with this large city and other smaller towns by regular bus service. These villages, consisting of 550 to 2000 inhabitants, are home to mixed populations. Two of the villages are made up predominantly of caste Hindus, more than half of whom belong to designated backward and scheduled castes. They are also home to significant tribal as well as Muslim minorities. A third village is divided between one hamlet consisting of Banjara people (a low-status backward caste) and another hamlet consisting of Gond tribal people, while a fourth village consists entirely of Bhil tribals. The primary economic base is farming, with wheat, maize, and lentils being the major crops. Most landed households own under three hectares, and landlessness is estimated at 11 to 30 per cent of the population.

Teak and mixed forest is accessible to all villages for firewood collection and grazing of livestock. In addition, villagers collect various non-timber forest products such as *mahua* (*Madhuca indica*) flowers, fruits, fibers, and medicinal plants, some of which serve as alternative income sources for the poorer sectors of the population. Villagers commonly grow fruiting and flowering trees such as mango (*Mangifera indica*), guava (*Psidium guajava*), papaya (*Carica papaya*), custard apple (*Annona squamosa*), ber (*Zizyphus jujuba*), oleander (*Thevetia* spp.), and plumeria (*Plumeria acutifolia*), among others. In addition, they grow certain speciality trees such as eucalyptus (*Eucalyptus* spp.), bamboo (*Dendrocalamus* spp.), and sacred tree species like *pipal* (*Ficus religiosa*) and *tulsi*, or holy basil (*Ocimum sanctum*). Trees are grown in many places on the village landscape, but the most common place is in the areas surrounding or adjoining houses, which I will term 'home gardens'. These gardens vary widely in size, depending on the proximity of neighbouring houses, among other factors, but an average size of a sample of about 30 gardens was approximately 150 sq m. Many gardens, however, contain only a handful of large trees and a few smaller plants because space must be reserved for household activities. In addition, tree planting is limited by low water resources and the necessity to protect seedlings constantly from livestock.

LOCAL HISTORY AND INTRODUCTION OF EXOGENOUS INFORMATION

This area has a particularly interesting history from the perspective of the evolution of local knowledge and the importation of exotic technology. Prior to Indian Independence in 1947, this region comprised part of the State of Bhopal, one of the princely states closely affiliated with the Raj of British India but never directly controlled by it. The state was governed from the city of Bhopal by Muslim rulers.

The personal lifestyles and official policies of these rulers during the late nineteenth and early twentieth centuries reflected their admiration of English ways, especially with respect to education, science, architecture, and landscaping, with important ramifications for agriculture and horticulture. For example, the agricultural estate belonging to the last Nawab before Independence, situated only a few km from the study area, was the first farm in the area to use sprinkler irrigation technology, American, English, and German motorised farm equipment, and imported seeds. The Department of Agriculture under this Nawab also held frequent agricultural fairs in one of the study villages, with exhibits demonstrating various aspects of modern farming. In addition, the villages in the immediate locality felt a more personal influence of the Nawab since one of his country mansions, used as a hunting lodge, was built nearby. This lavish estate was filled with exotic fruiting and flowering trees and the best varieties of Indian fruits such as mango, with seeds and grafted saplings imported from great distances in India and abroad. Presently, older village residents can still recall these tree specimens with great admiration, and some individuals who had worked there as gardeners still remember techniques used to manage them. Understanding this history of contact with externally-generated technologies is crucial to understanding the context of indigenous knowledge in this particular area, a topic which will be addressed in a later section.

Today, government-sponsored information sources for farming still enter the villages in the form of agricultural extension services. Radio and television programmes also provide farming information from outside. In addition, a Forest Department nursery sells seedlings of fruit trees and exotics such as eucalyptus and acacia species. Farmers also purchase commercial fertilisers and pesticides at nearby towns and at an agricultural cooperative in a village.

Definitions and analytical approach to knowledge

As used here, the term 'indigenous knowledge' implies any knowledge produced by and/or distinctive to a particular culture group or any loosely defined group of resource users in a given area. Such knowledge is usually produced informally by people actually practising a craft or working directly with a resource, as opposed to paid knowledge professionals such as scientists. As such, the term is synonymous with 'folk knowledge'. As used here, however, the term can also be expanded to include formally studied indigenous sciences such as Ayurvedic medicine. Although 'traditional knowledge' is often a subset of indigenous knowledge, not all indigenous knowledge need be traditional as it can also consist of recent innovations by local people. 'Global knowledge', 'international science', and 'industrial science', on the other hand, are used interchangeably to refer to knowledge produced by professional scientists usually at universities, research labs, and so on, whose sole occupation is the production and dissemination of knowledge and who, in connection with this endeavour, participate in a global culture of scientific communication. These definitions and their implied boundaries are intended rather loosely, in recognition of the fact that it is often difficult to determine in the field where one knowledge system ends and the other begins (Brodt 1998). In addition, these definitions are not intended to imply any clear, generalisable differences in characteristics of the knowledge itself.

For the purposes of this study, all knowledge systems are viewed as composed of hierarchical levels of abstraction, ranging from concrete practices to abstract concepts, following Horton (1982), Ingold (1992) and Kalland (1994). The more basic, or 'primary' level, consists of pieces of information resulting from simple cognition of objects in the environment and their cause-and-effect, spatial, and temporal relationships. In a knowledge system about tree management, for example, this level includes the information that trees will be green and healthy if dried cow-dung is applied to the soil at certain times.

One or more 'secondary' or 'conceptual' levels consist of those overlying constructs that are abstracted from the information level and that explain and unify the information pieces into a meaningful whole, the knowledge system. In Horton's description, secondary levels work by postulating a 'hidden realm' of entities and processes of which the experiences provided by the primary level are the surface manifestations. Concepts can include the idea that cow-dung contains particular nutrients essential for plant growth, or that 'hot' and 'cold' must be balanced within living organisms in order to maintain health. Whereas the primary level is closely tied to physical reality, conceptual levels are one step removed from the physical. This conceptual framework of knowledge assists in understanding why some pieces of knowledge may have different probabilities of persistence over time than other pieces.

Socio-economic factors affecting the communication of knowledge

Ultimately, for knowledge of any type to survive over time, it must be communicated among people and from one generation to another. Under current globalising trends, the powerful communication technologies employed by proponents of international science and technology have managed to disseminate their information even to the remote villages of India. It is this global communication that concerns scholars of development and culture, who warn that as people become increasingly integrated into the global economy, both materially and informationally, dissemination of their own indigenous cultures and knowledges declines (*see* Howes 1996). The implication therefore is that socio-economic factors that favour integration into a world system *disfavour* indigenous knowledge.

Many social and economic factors influence the tendency of different types of agricultural knowledge to be disseminated. High-income farmers throughout India lead very different lives from low-income farmers, and these differences greatly affect their abilities and tendencies to access and utilise knowledge of different types. These differences are best illustrated by a few representative case studies of members of different socio-economic groups.

ELITE FARMERS

Among the wealthiest landowners in this area are Muslim descendants of the ruling classes of the State of Bhopal who inherited up to several hundreds of hectares each of some of the best farm land. Ahmed (pseudonym), with whom I became acquainted, seemed typical of this class. He has frequent, often personal, contact with commercial entities such as fertiliser manufacturers, who provide information on the latest products. In addition, Ahmed occasionally contracts with Bhopal-based seed companies to multiply seeds of new crop varieties. He therefore plays a direct role in the perpetuation and dissemination of modern crop technology. He was also one of the first in the area to use a large combine for harvesting. In addition, he obtained grafted saplings for his extensive fruit orchard from private, out-of-state nurseries. Finally, like his distant relative, the last Nawab of Bhopal, he is a progressive thinker highly conversant in global agricultural trends and environmentalist thinking which he has acquired *via* a university education, international travel, and the global television and print media. As a result, he has experimented with commercial organic fertilisers, with which even the local extension agent is unfamiliar.

The knowledge base of this class of elite 'gentlemen farmers' reflects their extensive communicative capacities. They are generally less conversant with some of the more purely indigenous conceptions, such as a humoral 'hot'/'cold' classification of soil inputs, and certainly do not ascribe to ideas of sacredness of certain tree species, common in Hindu culture. They do, however, utilise some traditional and indigenous practices, such as the application of cow-dung as fertiliser, frequently in combination with elements of industrial agriculture. On the other hand, they are better able to discuss topics such as soil chemistry, using some standard university terminology.

Their overall knowledge framework, therefore, seems to be dominated by that of international industrial science, although some aspects of a more indigenous science are also evident.

MID-LEVEL FARMERS

With very little or no formal education, the situation for people at middle and lower socio-economic levels is quite different. They cannot use print media, nor can many afford radios or televisions. In addition, the overwhelming majority of those interviewed reported having very little or no contact with the agricultural extension system, Forest Department, or other sources of external information. Large seed and chemical companies do not bother to advertise directly to farmers with only small plots of land who probably cannot afford most of their products anyway. Instead, these farmers must rely on knowledge gleaned mostly from personal experience, relatives, and neighbours. It should be noted, however, that the more average mid-sized and mid-status farmers have somewhat more opportunities for learning and applying externally-accessed information than the very poorest and most marginalised sectors of the local population.

Sita Ram Rajput (pseudonym), for instance, farms just two hectares of land and has no formal schooling but, as a Rajput, he belongs to one of the predominate castes in this area. Therefore, while he does not have any direct contact with extension agents, he is well-connected in the community and can consult with other literate and more well-to-do Rajputs, whom he considers to be good farmers. He also listens to a regular radio programme on farming. Many in his class are able to purchase some chemical pesticides and fertilisers for their crop fields, although they also apply cow-dung. Many maintain impressive fruit trees in home gardens, most of which were planted from seed acquired in fruits that were eaten. They may occasionally purchase some trees, such as eucalyptus and less common fruit trees, from nurseries.

Therefore, while hovering on the edges of industrial-era technologies and international science, most of the average villagers like Sita Ram reported that the bulk of their knowledge about trees and crops came from their own experiences while working in their fields and gardens. Relatively little is acquired from outside the immediate family, and communication with those of very different social standing is limited and ritualised. For instance, many farmers make reference to the wealthy Muslim landowners as people to watch, since they are known to bring in novel techniques. Direct, open communication with them, however, is generally limited. When it does occur, it frequently follows set patterns in which the higher-status participants dominate the conversation while lower-status parties remain subservient and merely answer questions. The debt relations that commonly exist between the rich and the poor further reinforce such uneven patterns of interaction. Such conditions do not facilitate equal exchange of new information and ideas.

Indeed, the overall knowledge framework of Sita Ram and his peers has a predominantly localised and indigenous character: these farmers discuss the water-holding capacities of yellow versus black soil, hot and cold qualities of soil inputs, and the uses of soil inputs available within the village such as fish and buttermilk (the latter occasionally used for 'cooling' soil during the hot season). Villagers may fit individual elements from industrial agriculture, such as manufactured fertilisers, into this framework (Brodt 1998), but many do so without fundamentally disturbing the overall conceptual system that already exists. For example, they describe commercially available urea, a nitrogenous fertiliser, as being very 'hot' and requiring moderation in application, while cow-dung is relatively much cooler. In contrast, while the wealthy Muslim farmers might agree that urea must not be applied in excess, the designations of 'hot' and 'cold' are not used by them to describe soil inputs.

On the other hand, losses of knowledge attributable to market integration are evident for poorer members of this middle group. This phenomenon is best illustrated by the loss of knowledge about many forest foods, especially certain roots, among the younger generations of Gond tribals in the study area. While several older men could describe at least half a dozen or more edible forest plants that they knew from their childhood, and which they could confirm still exist in the surrounding forests, younger men reported difficulty in recognising them, even though they knew of them. Several decades ago this area was more isolated, with poor roads, fewer villages, and a smaller population overall, and people had to travel longer to reach the markets. Farmers also grew more traditional varieties of crops which reportedly did not yield as well as modern varieties. Consequently, the tribal villages that existed in this area had to be more self-sufficient than they are now, and forest foods were important as dietary supplements. Currently, however, these foods have lost their significance in the local diet now that other sources of food, including packaged snacks, have become accessible in nearby markets and village shops. Economics and market access alone, however, may not be the only factor in this loss of knowledge, since eating such foods from the forest is generally considered by all villagers to be something only tribals do. Such a designation might imply that forest foods are considered lower-status and perhaps undesirable by mainstream rural culture (*see* Daniggelis 1997 for the low status of wild foods in Nepal). Since the tribal people in this area are fairly well-integrated into mainstream Hindu culture, compared with the more remote tribal populations in eastern Madhya Pradesh, one could speculate that they themselves prefer not to be associated with forest foods. Such sentiments, however, were never directly expressed or hinted at to me (issues of prestige are addressed in more detail below), and a member of the older generation, in fact, expressed nostalgia for such foods. Therefore, other factors such as difficulty of collecting and processing these foods relative to mainstream crop foods may also affect food preferences. At any rate, the decline in economic need for these plants has precipitated a decline in indigenous knowledge about them, even though the plants themselves still exist in the landscape.

THE POOREST VILLAGERS

Finally, the poorest of the poor, those who live socially and economically on the margins of mainstream society, have the least access to outside knowledge, even from other villagers not of their class and social standing. One landless Gond (a scheduled tribe) family living on the outskirts of a village of predominantly caste Hindus, for example, must eke out a living from irregular agricultural labour and a few livestock that can be grazed on common village land or in the forest. Due to stricter forest regulations, family members are prevented from earning money by collecting firewood and selling it in nearby towns as their forefathers had done. Every member of the family is illiterate and, given their low social status, they appeared unlikely to associate much with the dominate Rajput group in their village. Their opportunities for acquiring new information are thus generally limited to their experiences of working on the farms of wealthier people. Confined to the small space around their house they also lack the capacity to grow many trees or other plants through which they could develop their own knowledge base. Their familiarity with both international science as well as some of the prevailing indigenous concepts such as those involving soil types and hot and cold inputs is therefore limited.

These three case studies illustrate the large range of different learning environments inhabited by different classes of villagers. Certainly, they do not encompass all individuals. Some innovative but relatively poor individuals in this study, for example, were able to glean much information from different employment experiences on large farms and develop it further in their own small home gardens despite their overall paucity of educational and material resources. And higher-class women, in general, often lack the informational resources that their husbands and even lower-class women enjoy, since they tend to remain sequestered in *purdah* (seclusion of married women prevalent in Hindu culture) inside the house and are not involved at all in farm work. Since the labour of lower-class women is often required in the fields, they cannot observe *purdah* as strictly as upper-class women and thus may encounter more opportunities for learning outside.

Some generalisations can nevertheless be drawn from the three broad categories outlined above. Neither the most elite farmers nor the poorest villagers are likely to have the richest supplies of purely 'indigenous knowledge' of farming and tree cultivation. The elite rich are too heavily invested in industrial technology and the global information streams that accompany them. While this investment does not preclude the necessity and creativity to adapt information to their own localities, it does mean that they do not feel much of a need for some of the more commonly known indigenous or local concepts and practices. Scale seems to be an important deciding factor. With as much as 50 hectares of productive farmland, Ahmed noted that he would not be able to procure enough local materials, even cow-dung, to suffice for all his fields. Large quantities of commercial inputs, on the other hand, can easily be purchased in the market and transported to his farm. These findings, therefore, support scholars' concerns about the detrimental impact on indigenous science of easy access to a global system.

The lack of access of the poorest classes to international science is not necessarily, however, complemented by a wealth of indigenous scientific information. The socio-economic factors that limit access to external information are often the same factors that limit learning of the community's indigenous knowledge or development of their own experiential knowledge. The idea that indigenous knowledge is richest in those populations least touched by industrial technology or the global market may be correct in more remote communities that have access to natural resources and can make a living practising traditional crafts outside of the market system (Gupta 1997). It may, however, be erroneous in heterogeneous, market-integrated communities where being 'scientifically untouched' means being socially and economically marginalised even from local resources required to build indigenous knowledge, especially technical knowledge of crop and tree cultivation, which require access to land.

Rather, those with the greatest store of indigenous technical farming knowledge are villagers in the very broad and diverse middle category. While they have some access to external information, they operate on smaller scales more amenable to the use of local materials. In addition, their contact with global systems is not intimate or frequent enough to facilitate transfer of many of the more subtle components of knowledge such as abstract secondary-level conceptions of how and why things work as they do. One should note, however, that these people's knowledge is by no means uniform, nor is it easy to distinguish any sort of 'pure' indigenous knowledge. Because individuals have varying degrees of access to outside scientific information, their personal knowledge bases exhibit a wide variation of mixtures of two or more recognisable knowledge systems, with each person possessing a slightly different combination of techniques and concepts (*see* Brodt 1998).

CASTE HETEROGENEITY AND KNOWLEDGE DIVERSITY

The illustrations above suggest that economic and ethnic heterogeneity of a population foster a diversity of knowledge. Heterogeneity is significant partly because of 'homophily' in communication, a concept propounded in the innovation diffusion literature (Rogers 1995; Gogoi 1990). Homophily is the degree to which two people who communicate are similar in a given set of characteristics, which may include socio-economic characteristics. One of the principles of communication theory is that communication is more frequent and effective between homophilous than between heterophilous individuals. If communication of knowledge *between* socio-economic groups is not as high as *within* groups, then the resulting knowledge of the population as a whole is likely to be more diverse.

In India, however, simple socio-economic homophily of communication may not suffice to explain the sheer diversity of lifeways and accompanying knowledge systems. The variable of caste adds an additional criterion of identity which impacts the conservation of knowledge, as castes (in the sense of *jati*) were traditionally based to a large extent (if not solely) on occupational differentiation. Different castes specialised not only in doing different hereditary jobs, but also in utilising different natural resource bases. On one island in Karnataka, for instance, there are two castes specialising in mat weaving, but only one caste utilises *Pandanus* leaves while the other uses only *Cyperus* reeds. Different hunting castes hunt different prey species, and so on (Gadgil and Guha 1992). The present study included members of the Basor caste, which traditionally specialises in bamboo weaving. Members of this caste have knowledge about the utility of different species of bamboo and how to store bamboo poles underground to prevent them from rotting.

This connection of natural resource-based knowledge with caste has key implications for conservation of knowledge diversity. Caste in India is intimately related to social and personal identity; even when people change occupations, they still retain their caste membership (Beteille 1996; Kangayappan 1992). Moreover, despite official attempts to eradicate distinctions of caste due to the traditional inequities associated with them, certain forces in modern society, such as the electoral process, urbanisation, and labour movements, are actually strengthening traditional caste ties (Beteille 1996; Sinha 1995). Caste, therefore, serves as a convenient and important source of personal identity, even in the face of shifting social status and occupational roles, and it is not something easily relinquished. In so far as specialised resource-related knowledge is tied to certain castes, then, such knowledge itself may serve as an important identity marker not readily abandoned. This proposition is supported by the resurgence of interest among indigenous peoples around the world in reviving traditional crafts and accompanying resource management techniques as a way of asserting their unique cultural identities (Friedman 1989). The large degree of social and ethnic heterogeneity inherent in Indian society might therefore aid in preserving diversity in indigenous knowledge.

Hidden landscapes

As described above for the mid-level farmers, most people mix knowledge and techniques from different sources, depending on their degree of access to externally-derived science. The result of this mixing is that any given field or garden might have several different practices and concepts applied to it, some with a more local and some others with a more external origin. Nevertheless, a few areas of the landscape seemed to retain a slightly more 'indigenous' character than others in terms of the type of knowledge applied to them. Home gardens as a group, for instance, are often more highly associated with what one might consider to be indigenous knowledge than other areas of the landscape, especially crop fields. As such, they may function more effectively as repositories of such knowledge and thereby further confound any simple associations of global integration with decline of indigenous knowledge.

The reasons for this spatial association are multiple. Generally less attention and effort is invested in gardens than in crop fields, probably because gardens are usually kept for subsistence and ritual purposes rather than economic gain, and what foods they produce are not considered staple foods. Even the maize that is often grown in home gardens during the rainy season is eaten mostly as a fresh vegetable snack, rather than as a staple grain. In some cases a home garden is planted and cared for primarily by one person (who may be male or female) in the family who tends it as a 'hobby' (*shauk se*). In concert with the lesser economic importance of gardens, therefore, villagers seem to place lower priority on obtaining the most current management information for them, especially from outside the village. Moreover, since external inputs such as fertilisers and pesticides tend to be expensive for most households, they are generally eschewed in home gardens and applied only on crop fields, if at all. Issues of scale are important here also — households are more likely to be able to generate local inputs such as cow-dung, ashes, and buttermilk for their smaller home gardens than for entire crop fields, which are usually many times larger.

In addition, these gardens are conveniently integrated into outdoor living and work spaces close to the home and require relatively small amounts of year-round labour inputs that can be timed flexibly. All of these factors allow labour in the home garden to be shared by many people (except where it is dominated by one or two individuals as a hobby). More significantly, even women and children of better-off households who are not required to work in the fields can care for the gardens in their spare time between other household chores. This pattern of management contributes to the paucity of external knowledge applied to home gardens, since women and children are less likely to have contact with formal external information sources and materials.

Indicative also is the fact that there is no local equivalent to the English term 'home garden'. The only way to refer to this area is to talk about 'trees planted by the house', 'behind the house', etc. The Hindi term *bagiichaa*, often loosely translated as 'garden', is in fact used solely to refer to large-scale orchards planted only by the wealthiest villagers. In addition, some home gardens are actually *khalihaan*, large fenced areas, usually close to the village, kept empty in the centre for crop threshing, cleaning and temporary post-harvest storage. When trees and other plants are planted around the protected inside edges of a *khalihaan*, it closely resembles a home garden. Ironically, while the term 'home garden' does not have a clear local translation, the term *khalihaan* does not have a clear English equivalent.

The namelessness of these gardens accentuates a certain air of official invisibility. Compared to crop fields and forests, they remain outside of the official limelight which might otherwise bring more exogenous knowledge to bear on them. When forest officers and the agricultural extension agent were first asked about villagers' tree planting practices, none of them initially acknowledged that villagers planted trees at all, thereby completely neglecting home gardens where most trees are planted. These gardens and the plants in them are therefore hardly ever visited by outside experts, and the agricultural agent, indeed, personally confirmed that he seldom gives advice about fruit trees even though they are officially within his mandate. This invisibility is confirmed by Dove (1992), who noted a similar administrative invisibility of non-marketed trees in Pakistan.

By extension, therefore, anything that is grown around the home and in *khalihaan* is a potential repository of indigenous knowledge. Since trees, especially the more valuable fruit trees, are grown much more often in home gardens and *khalihaan*, where they can best be guarded against livestock and theft and be watered more easily than in other places, they automatically fall into this category. For example, applications of buttermilk and cow-dung and pruning of roots, all indigenously-developed technologies, are generally done in the home garden or on trees planted on fields, but not on crops on the fields. Although the problem of scale — most people only have enough buttermilk at any one time to suffice for a few trees and not for an entire crop field — is one important reason for this distinction, it may also arise out of the lesser overall significance of trees and gardens. For instance, several people explicitly mentioned that urea is only meant for crops and not for trees, implying that it is considered inappropriate for trees. Such perceptions may arise if vendors of commercial fertilisers or other outside information sources decline to recommend such products as suitable to trees, perhaps due to their ignorance of home gardens. In addition, planting material is rarely purchased for home gardens, where trees are generally grown from seeds procured directly from fruits or from cuttings taken from other trees. In contrast, purchasing new improved or hybrid seeds for crops is much more common.

This invisibility of gardens and their associated knowledge holds important implications for the survival of indigenous knowledge. As places where people can apply and experiment with knowledge relatively undisturbed, gardens serve as potential living libraries of indigenous knowledge of plant cultivation. In so far as crop-related and tree-related knowledge overlap, this characteristic of gardens may benefit crop indigenous knowledge as well. As the economic valuations of landscapes change over time, however, this role of trees and home gardens may change as well. For example, a rapidly increasing number of households have recently been able to install electric tube-wells on their fields, which enable not only crop irrigation but also more tree planting on fields. If the future were to bring an increasing emphasis on more prestigious orchards located separate from the house, which only the very rich have now, then farmers might begin to seek out more external knowledge and technologies to cultivate them. Such a change in the landscape might also create greater divisions between the knowledge held by upper class farmers and that held by lower classes who cannot afford to plant orchards, or between men who work out in the orchards and women who stay at home.

Such a scenario would also provide more opportunities for the labouring classes to acquire external knowledge through employment in other people's orchards. Such opportunities have already come and gone at the Nawab's local estate. Many elderly, scheduled caste men in this study could describe exotic fruits and ornamental shrubs that they cared for as youths while employed as gardeners or labourers at the estate. These plants, however, are relatively unfamiliar to their children and grandchildren now that the estate is in decay and the plants have all but disappeared. In this case, economic and physical changes in the landscape prompted a loss of externally-derived knowledge.

Evidence suggests, however, that the reverse scenario, of a decline in indigenous knowledge due to more orchard planting away from home gardens, may never actually come about. Even in the face of large-scale landscape change, some niche for indigenous knowledge may persist. One can imagine that some elderly grandmother or even some new young wife may want to grow a fruit tree or two and some flowers for religious rituals, without incurring much expense to the household or taking much time from other family members. Even at present, one elderly family matriarch grows flowers, vegetables, and some fruit trees behind her kitchen as a hobby without any external inputs in spite of the large, immaculately managed, chemically-fertilised orchard of expensive grafted fruit trees owned by her family just outside of the village. It is in such hidden interstices of hobby kitchen gardens that indigenous knowledge may survive.

In a similar vein, evidence of persistence of traditional crops, even under conditions of economic and technological change, is available from other parts of the world (e.g., Richards 1985). In the Andes of Peru, Brush (1992) describes how even farmers who are highly dependent on commercial sale of improved varieties of potatoes still manage to reserve some space on their fields for native varieties. In fact, they can scarcely imagine any reasons for not growing some native potatoes, since these varieties fill important culinary and social niches (as 'gift potatoes') which the improved potatoes, by their very nature (they are considered to be more insipid and watery), are unable to fill now and perhaps ever in the future.

Perceptions as factors in communication

PERCEPTIONS OF KNOWLEDGE

The observation that socio-economic factors affect communication patterns, while useful, is ultimately in itself not an adequate explanation for the postulated demise of indigenous knowledge when communities become globally integrated. The mere fact of access to outside information and materials does not require one to reject one's own community's knowledge traditions. Scholars, however, identify people's perceptions of knowledge and its sources as key factors in the conservation or decline of indigenous knowledge. The significance of perceptions and prestige in adopting new knowledge and technology is recognised by scholars in the field of innovation diffusion (Rogers 1995). In the indigenous knowledge literature as well, some types of knowledge (i.e., global scientific knowledge) are thought to carry more social prestige within local communities than other types (i.e., indigenous knowledge). Thrupp (1989), for instance, describes small-scale farmers in many countries as being all too willing to abandon traditional folk wisdom in the face of the perceived superiority of international, mainstream science. Such perceptions are encouraged, wittingly or unwittingly, by development projects and extension systems that tout the power and modernity of institute-developed, high-technology approaches to farming. This type of development can leave local people feeling embarrassed about their 'backward ways' and eager to adopt the knowledge of higher-status extension agents, according to Thrupp (1989) and Howes and Chambers (1980).

While issues of prestige do pose a serious threat to indigenous knowledge systems, findings here suggest that these issues are often just as complex as socio-economic factors, to which they are tied. First, in order to compare different levels of prestige of different knowledge systems, one must be able to distinguish the knowledge systems themselves as distinct from one another. In this study area, however, people generally did not distinguish, either verbally or actively in their management activities, between different types of knowledge systems such as outsider's versus insider's information, or modern versus traditional knowledge. Only two or three of the villagers interviewed indicated that nowadays mostly 'new' farming techniques were being used, such as chemical fertilisers and commercial seed varieties, as opposed to 'old' techniques such as cow-dung and traditional, self-saved crop varieties. The only person who made any distinction based on locality was a highly educated manager of a large absentee-owned farm. This man labelled some of the villagers' practices with which he himself was unfamiliar as *deshii*, meaning country or rural. His use of this term, however, did not seem to be necessarily pejorative, but rather intended to imply that he understood that people who do not have access to higher-level city education must create their own solutions to problems where they are, in the country.

When it came specifically to trees, however, the great majority of people easily mixed exotic species with native species without a second thought, and spoke about urea (using the English word) and cowing in the same breath, without making any distinctions as to which was 'new' or 'traditional', or native or outside. The lack of attention given to spatial distinctions of knowledge (indigenous versus external), suggests that either people are not aware that knowledge elements are coming from different places, or that such differences are simply taken for granted as normal to the constitution of local knowledge. A further implication is that the introduction of knowledge from outside the immediate locality is not a new phenomenon, but is a process that has already been occurring for a long time. Such, indeed, has been the case considering the historical influence of the Nawab in promoting new agricultural technologies, as described above.

These observations lead to some crucial insights pertaining to knowledge conservation. First, in communities that have not historically been completely isolated socially and economically, the joining of outside knowledge with the already-existing indigenous knowledge can be considered a usual and normal phenomenon. Furthermore, the current absence of a specific development intervention in agriculture in the study area as well as most people's lack of personal contact with the extension system suggests that local people may feel relatively less pressured by outside forces to adopt new technologies. The presence of either of these two factors alone bodes well for indigenous knowledge conservation. Together they positively reinforce each other; as long as the process of incorporating outside information is a normal fact of life, local society as a whole is likely to be more accustomed to the process of weighing new information and considering it on merits besides just prestige. Consequently, local people might be somewhat more immune to the more detrimental aspects of a development intervention, such as hasty, ill-planned technology adoptions, if such an intervention were to begin in this area. On the other hand, the current absence of a development programme means that when new ideas and technologies do enter the area, they possibly do so in a less pressured way, through local contacts, allowing people more leeway and time in deciding for themselves if and how to incorporate such newcomers.

PERCEPTIONS OF KNOWLEDGE SOURCES

Even when different types of knowledge *per se* are not thought to convey different degrees of prestige, the sources from which knowledge is disseminated might carry differences in status. Such status differences affect communication patterns because a person actively seeking knowledge generally does so among people whom he or she believes to be more knowledgeable relative to himself or herself. And people of higher status are often thought by others to be more knowledgeable; whether their higher status is due to their actual knowledgeability or other socio-economic factors. Therefore, understanding perceptions of status and how they are tied to perceptions of knowledgeability can be important in analysing information-seeking behaviour (*see* also Rani and Malaviya 1991).

Individuals within the study area expressed definite ideas about the people believed to possess the best knowledge about farming and tree care. When asked for the names of the most knowledgeable individuals, interview respondents commonly named one or two of the biggest, wealthiest farmers in the village or local area, such as the elite Muslim landowners described above. When questioned further, all other villagers could be lumped together as being equally knowledgeable. This correlation of knowledge with wealth suggests two possible interpretations of villagers' perceptions. One possibility is that wealthier farmers, with higher education and better communication resources, are perceived as having more access to outside knowledge, which is considered valuable. The second possibility is that knowledge is highly associated with physical evidence and is meaningless in the abstract. Therefore, those who have the necessary material resources to manifest their knowledge, such as by establishing impressive-looking mango orchards, are automatically respected for their evident knowledge. Potential or latent knowledge of a person who lacks the necessary material resources to put it into practice is not perceptible and therefore either unknown or not considered worthy of mention.

The former proposition that outside or formal, global scientific knowledge is considered valuable was particularly espoused by those with close ties to educated people. For instance, a respected brahmana man with little formal education but with a significant repertoire of local knowledge of tree and crop care strongly recommended the knowledgeability of another man with whom he had a close working relationship. He noted that he specifically chose this individual due to this man's formal education (having attained a Master of Science degree in agriculture). In other words, while himself possessing a respectable store of informally acquired knowledge, he chose someone unlike himself, with much more formal education, to represent his conception of knowledgeability. When pressed further, he also recommended a good friend of his who maintains close contact with extension services and runs a high-technology and high-input operation, while retaining significant amounts of local folk knowledge as well.

A few other farmers also supported the value of formal education in increasing knowledge. For example, one illiterate man specifically noted that he consults literate farmers with much land for advice on agricultural matters. One woman gave a rather fatalistic account of village tree planting behaviour, expressing her view that villagers just plant something without giving much thought to where or how it is planted, and whether the tree subsequently flourishes or dies. Thinking ahead of time about soil requirements or other planting factors, according to her, was 'city talk' (*shaher ki baat*), implying in her explanation that only educated city people, who are by definition considered more knowledgeable than villagers, would give some thought to their planting procedures.

Wealth and formal education, therefore, both appeared to be high-prestige characteristics that endow individuals with knowledgeability in the eyes of their neighbours. This conclusion on the surface seems to threaten the survival of indigenous knowledge, as formal education is generally depicted by scholars as erasing or, at best, ignoring, local forms of knowledge (Clarke 1990; Eisemon 1989; Flavier *et al.* 1995; Gupta 1998). One is also tempted to conclude further that if community members as a group tend to admire and want to emulate the elite and educated among them, then their efforts to perpetuate other forms of knowledge may wane. Such a conclusion, however, would be overly simplistic in the absence of detailed evidence to support it. First, while the average villager may wish to be like the educated elites, he probably is not actually able to be like them due to lack of material and informational resources. And lacking their extensive access to global knowledge, the average villager will of necessity continue to work on his store of locally-generated information. Second, wealthy elites, while referenced often by their neighbours, were not the only people seen as knowledgeable. Therefore, wealth and education, while important, are not necessarily the only indicators of knowledgeability used by local people.

Certain other individuals who were not wealthy or highly-educated but who did have good home gardens were also occasionally named as knowledgeable. This observation supports the second hypothesis proposed above that knowledge must be proven by physical evidence to be respected by other people. In all cases in which a villager with very little or no land was referenced by others as having good knowledge about trees, the person referred to at least had a distinctive home garden as visible evidence of his tree knowledge. People who did not have any distinctive tree plantings but who had, for instance, worked for many years as gardeners at the Nawab's estate or who otherwise demonstrated considerable knowledge in interviews were never cited by other villagers. Den Biggelaar (1996) found a similar correlation between physical evidence and perceptions of knowledge in a study of agroforestry knowledge in Rwanda. He found that people only recognised a person as knowledgeable if he or she revealed his or her knowledge through action. Therefore, "knowledge was defined primarily as having species diversity on the farm" (den Biggelaar 1996: 66).

In addition, some mid-level farmers noted that they do not specifically consult wealthy farmers on farm-related problems because of a perception that the wealthy farmers have to face the same problems and consult the same extension agents as poorer farmers, thereby showing that they have no more expertise than anyone else. In a small group discussion about farming, two men specifically explained that resource-poor farmers have the same information as wealthy farmers, but they just cannot utilise it as effectively because they do not have the necessary material facilities, such as irrigation equipment and chemical fertilisers. These same farmers expressed a belief that the real-life experience and practical knowledge gathered by elderly people over time is more valuable than the greater formal schooling of young people. They noted that this valuation applied especially in the local area, where only up to a tenth grade schooling was available to most young people. The discussants did not think that this amount of schooling conferred any major advantage for obtaining significantly better quality information than was available to less educated older people. Such an assertion implies that the farmers believed that valuable knowledge and information may indeed exist somewhere outside of the local community for people with a very high degree of education. However, that which is easily and locally accessible to someone with only a moderate degree of education is not of any greater value than the more traditional, time-tested knowledge of older people. This statement does not necessarily conflict with statements by other people noted above about the value of education, since most of these latter were referring to a 'city education' or a university education, not available locally. It also supports the idea forwarded by many scholars that formal schooling does not help people to acquire local indigenous farming information, although whether or not it actually hinders the acquisition of such knowledge by devaluing it was not clearly addressed.

While some villagers may see formal education as beneficial for developing farming knowledge, not all formally institutionalised information sources are considered equally useful. One might expect that the Forest Department, for example, would be considered a common source of information about trees, especially since the Department nursery sells bamboo, eucalyptus, and fruit trees to local villagers and Department staff have pursued formal courses of study in forestry. It is also much more accessible to average villagers, since lower-ranking staff live directly in the villages and a range office is located in one of the study villages. Nevertheless, villagers interviewed overwhelmingly felt that Forest Department staff were not more knowledgeable than villagers about growing tree species suitable for home gardens, and most villagers do not consult with Department staff about their trees, even when they purchase seedlings from the nursery. In fact, the Forest Department was the least accessed external information source among a sample of diverse individuals. Most people indicated the reason for their views as being that the Forest Department's job is to protect forest areas, not to grow trees, and that therefore, compared to experienced village people, it does not have any additional expertise in growing trees. The only exceptions were a few individuals who thought that Forest Department staff had more knowledge specifically about forest species and exotics such as eucalyptus. No one, however, thought that the Forest Department has better knowledge overall, for all trees.

On the other hand, authority figures' perceptions of villagers' knowledge are also thought of as instrumental in fostering either the demise or the continuation of indigenous knowledge, as their attitudes favouring or disfavouring particular forms of knowledge may be transmitted to local people (Howes and Chambers 1980; Thrupp 1989). Local Forest Department staff were interviewed to assess their perceptions of villagers' knowledge related to tree planting, with mixed results. One individual at each level, from Divisional Forest Officer, to Forester, down to Forest Guard, expressed a high opinion of villagers' skill in tree planting, and the Forest Guard even maintained that villagers are much more knowledgeable about the local forest areas than Forest Department personnel. The Forester also asserted that he sometimes consults elderly villagers for advice on plantations because he respects their knowledge, even though he is aware that many of their techniques of tree care are different from what is taught in forestry training courses. One forest officer, however, said that villagers do not know much about forest management. Both Divisional Forest Officers interviewed also mentioned that one of the Forest Department's functions is to extend technical information to local people. Therefore, while they avoided outright negation of villagers' knowledge, they seemed to be operating under the assumption that the Forest Department has knowledge which villagers do not have. Overall, therefore, with one exception, Forest Department perceptions seem to parallel those of villagers, with villagers seen as being knowledgeable in some realms of tree planting, but with the Forest Department seen as having much better knowledge of forest management techniques. It is noteworthy, however, that at least some forest officers appeared to see some relevance of villagers' knowledge to forest management. The usual generalisations about official ignorance of indigenous knowledge therefore do not necessarily hold up in this particular case, especially at lower ranks, where staff have the most contact with villagers. In so far as these ranks have the least decision-making power, however, overall institutional ignorance, especially at higher levels, may still be a viable and urgent concern (*see* Dove 1992). Its direct assessment, however, was beyond the scope of this study.

Cognitive characteristics of knowledge systems and knowledge users

Concern over knowledge loss often conjures up an image of one set of practices and concepts simply replacing another set wholesale. Evidence here suggests, however, that such complete replacement and loss is uncommon. Certain characteristics of the knowledge itself and how it is conceived of by people can help parts of a knowledge system to survive even as parts of other knowledge systems assume common usage. These characteristics include the generality versus specificity of particular pieces of knowledge and the flexibility of the whole system.

Practices and concepts that are generally applicable to many different species are more likely to persist and become widespread in the population than those that are narrowly applicable to only a few species. For example, the practice of root pruning to induce flowering and fruiting, described by one old priest, seems mostly applicable to citrus trees, which not all households grow. The specificity of the practice, therefore, may be one reason why few other people seemed familiar with it when questioned.

On the other hand, a very generally applicable concept, such as the hot/cold dichotomy, is familiar to many and is being learned very thoroughly by members of the youngest generation. This hot/cold idea is not simply a way of describing the effects of cowing, urea fertiliser, buttermilk, and other substances when applied to trees, but also implies a much more general meaning. Heat is associated with potency or energy and when too strong can cause 'burning' and drying. When mediated by cold, heat can result in good colour and development of living organisms such as plants. Because of their general connotations, these terms are applicable to other domains of life in addition to agriculture. In the *Ayurveda* system of medicine, for example, drugs and foods all have *virya*, also described as potency, which can be either hot or cold (Dash and Junius 1983, Jolly 1977). These concepts are widely used in folk health knowledge throughout India, even by formally educated urbanites, who describe foods as having heating or cooling effects on the body. Therefore, one may assume that even landless families lacking any space for tree or crop cultivation may retain this concept quite securely in their overall knowledge systems. This widespread applicability, thus, will help ensure the conservation of this piece of knowledge over a longer timespan, even in the face of socio-economic and technological changes.

Flexibility of an entire knowledge system can also influence its chances of survival. From lessons learned in general systems studies (Allen and Starr 1982; Holling 1986; May 1973), we can postulate that knowledge systems with relatively independent sub-sections may survive better in the face of change than very highly integrated knowledge systems. In the study area, for example, the religious aspect of certain sacred tree species seems to function as a distinct sub-system of knowledge that is less likely to be eradicated in the face of technological modernisation. Some species, especially *pipal* (*Ficus religiosa*) and *bar* (*Ficus benghalensis*) are thought to be inhabited by deities, and they are therefore protected wherever they grow and are frequently planted in temples. While such trees have an important link with agricultural knowledge — most farmers understand their negative impact on crop growth due to soil, water and nutrient competition and their interference with farm machinery when located on fields — their religious meaning gives these trees a crucial independence from purely technical agricultural knowledge. Indeed, many farmers' fields, even those farmed with tractors and combines, contained sometimes quite large islands of sacred *Ficus* species. Although some villagers noted that there was an increasing tendency to cut these trees down to make way for new farm machinery, such trees and the sites they occupy still retain a large significance in the religious lives of villagers. This significance is attested by the recent construction of several new worship platforms around such trees even on mechanised farmers' fields, their prevalence at temples, and the recent plantings of such trees by people of many different castes and classes. These observations suggest that technological change does not affect the fundamental religious values associated with these trees. These values together with the knowledge of how to propagate the trees therefore form a relatively independent and stable sub-system of indigenous knowledge which is likely to persist in some form for a long time to come, even if the larger knowledge system becomes transformed or eroded. Furthermore, in so far as this sub-system encompasses practices and concepts also present in other areas of the larger knowledge system, the larger system itself may benefit from the independent significance of its sub-system.

The flexibility to create new linkages between elements of different knowledge systems may also be an advantage in conserving indigenous systems. For instance, farmers such as Sita Ram Rajput were able to establish a link between the traditional concept of hot and cold effects of substances and the new practice of applying commercial fertilisers such as urea to crops. By recognising urea as being 'hot', they effectively incorporate a new practice without simultaneously losing an already existing concept. In addition, the farmers conceive of urea as being a powerful nutritional substance, like 'food' for plants (the power of which is connected to its 'heat'), a concept similar to the idea of plant nutrition in international plant sciences. According to these conceptions, therefore, urea can be linked both to global sciences and to indigenous concepts simultaneously, forging a stable complex of ideas that reinforce rather than replace each other. Instead of disappearing with a change in material realities, an adaptable concept can acquire a new life when creative farmers are able to use it in conjunction with new technologies.

Finally, different levels of a knowledge system tend to have different survival potentials based on differences in their communicability. As shown by the many middle-level farmers who have incorporated exogenous fertilisers and other materials into their indigenous knowledge systems, concrete, observable practices such as fertiliser application tend to be more readily communicated and spread among an illiterate and highly heterogeneous population than the more abstract ideas about how and why they work. This observation may be especially applicable in South Asia, where, as described above, members of different subgroups of the population do not always communicate easily with each other on the deeper level required for transmission of more abstract ideas.

Conclusion and recommendations

I have only been able to touch briefly upon a few of the many factors that affect the durability of natural resource-related knowledge systems in India. These factors include opportunities for communication and learning, prestige of knowledge and knowledge sources, scale of operation, 'invisibility' of parts of the landscape, generality versus specificity of pieces of knowledge, and adaptability and communicability of the knowledge itself, in addition to personal factors such as individual creativity. These factors are very complex by themselves; furthermore, how they all interact with one another in any single community or in any given individual can be highly variable. For example, problems of scale and non-availability of certain materials can negate someone's personal interest in experimenting with more traditional techniques. Pride in one's ethnic culture or craft knowledge may conflict with economic needs for greater market integration. On the other hand, financial inability to afford certain materials can prevent adoption of more prestigious 'scientific' inputs and reinforce reliance on local materials. In addition, 'globalisation' of some members of a family through formal education and outside employment may allow other members more time and opportunity to continue growing plants using a more indigenous knowledge base.

These types of interactions between multiple forces within modern-day rural communities require further examination, for they can greatly complicate the simple equation that dictates that global integration must lead to the loss of indigenous and traditional knowledge. In a reversal of this equation, Bebbington (1996), for example, suggests that Andean villagers are able to preserve their traditional communities intact expressly because of externally-derived agricultural inputs that have boosted their otherwise meagre yields and made their communities more viable economically, thus preventing some of the emigration that is now so typical of mountain regions. By 'modernising' some of their agricultural practices, therefore, these communities are in a better position to preserve other aspects of their indigenous cultures. Brush (1992), as noted above, has also shown how commercialisation and modernisation of Peruvian farmers' potato crop has not erased a significant cultural niche for native potato varieties. These examples join those from the present study in showing that indigenous knowledge can and often does survive because it is useful and valuable in particular contexts and to particular groups of people.

On the other hand, warnings about the erosion of indigenous knowledge of natural resources should not be taken lightly, because such erosion is certainly occurring throughout the world. The analysis presented here, however, indicates that certain avenues are still open for moving toward conservation. Issues of poor prestige of indigenous knowledge are already being addressed by some organisations whose exemplary work should be expanded. The Society for Research and Initiatives for Sustainable Technologies and Institutions (SRISTI), a non-profit organisation in Ahmedabad, for example, organises biodiversity competitions among primary school children in villages to reward those children who are familiar with the most local plants and their uses. Such competitions may be judged by traditional healers and other knowledgeable elders from the local area. In addition, this organisation is working with the government of the State of Gujarat to establish a venture capital fund which will enable small farmers and craftspeople to apply for funds to develop their own innovations of small-scale technologies. Moreover, this organisation is calling for the government of India to devote more capital and personnel towards labs dedicated to adding value to local knowledge and combining the best of the formal and informal scientific sectors (Gupta 1996, 1998). Issues of granting intellectual property rights as a means of protecting and valuing indigenous knowledge are also being vigorously debated (e.g., Brush and Stabinsky 1996; Gupta 1996a; Singh and Mahanti 1997). In addition, many authors have noted that formal schooling in many countries tends to ignore indigenous knowledge systems and local innovation (Clarke 1990; Flavier *et al.* 1995; Gupta 1998), a failing that requires remedy. Moreover, Eisemon (1989) suggests that the teaching of global science in most African schools could be more effective if instructors began with knowledge their students already possess from social experience, i.e., indigenous knowledge, and use that knowledge as a way of introducing and bridging to new scientific concepts.

In addition, as propounded most notably by Chambers (1983), educated outsiders — i.e., high-status people such as development professionals and extension agents — should institute 'reversals in learning', in which they try to learn from local people. Such reversals can aid in valuing local knowledge and ensuring that it is not trampled by 'development projects' that do not account for indigenous modes of resource management. I would suggest that this may even entail purposely leaving some aspects of rural life, such as activities taking place in 'hidden landscapes', untouched by deliberate development activities, under the assumption that the continued use of indigenous materials may enhance the viability of the knowledge associated with them.

Another proposal suggested by the analysis presented here is to make indigenous materials more economically viable in a modern world. Issues of scale surface as particularly important in this area, with more people engaging in larger-scale operations which appear to negate the utility of local materials. For example, Ahmed, the wealthy local farmer, expressed some interest in applying more cowing to his fields as an organic fertiliser, but indicated that he could not procure the large amounts needed for all his fields very easily in the immediate vicinity, and transporting it by tractor from longer distances was not a viable option. The importance of scale in choosing between raw materials and their accompanying knowledge systems suggests that 'local' methods of farming (utilising more local materials) could be made more attractive if they were more cost-competitive for larger-scale farmers. Yet while chemical fertilisers and petroleum fuel are often the subjects of substantial government subsidies, cowing or the transport of wood ash is not. Policies that reverse this imbalance or at least eliminate certain advantages to the global scientific-industrial complex, while politically difficult, might enhance the attractiveness of indigenous methods. The growing linkages between some indigenous practices (albeit certainly not all) and modern concepts of 'sustainable resource management' may help to make such shifts more feasible. For people like Ahmed, for instance, the fact that cowing is an organic input might be more important than that it is a traditional and indigenous input. It is thus plausible that at least some types of indigenous knowledge may retain their value in new contexts and will therefore naturally be preserved by their users, perhaps even without extra promotion from outsiders.

Working to provide linkages between small-scale indigenous technologies and global economies is another important avenue to follow. Hill (1995), for example, describes how an engineer in Uttar Pradesh designed a new spindle that was small-scale and serviceable at the village level, but that produced a thread of the standard required for automated weaving looms. Prior to this invention, traditional cotton spinning in Uttar Pradesh villages was in imminent danger of demise because the existing spindles could not produce the quality of thread required by the new looms. By introducing one slight variation to the existing technology, the engineer enabled many villagers to continue their traditional home-based craft even as they were linked to global systems. In a similar vein, Ahmed, as noted above, was very interested in applying cowing to his fields through his seed drill. Unlike chemical fertilisers, however, the cowing clogged his machinery, and to apply it by hand when seeding required a prohibitive investment in labour for his large farm (besides the problem of availability noted above). Adapting a seed drill that could accommodate cowing might thus enable even large-scale commercial farmers to use this local material more extensively, and through continued use, the accompanying knowledge about the material may also be sustained.

New technologies, therefore, can sometimes allow indigenous knowledge to interface successfully with global technologies and economies. This point highlights the fact that indigenous knowledge is always changing and adapting to modern situations, and the more it can do so the better are its prospects of surviving in the long run. Gupta (1996) and SRISTI, the Ahmedabad organisation, clearly express this need for adaptation and creativity with their emphasis on grassroots innovations, and not merely traditional knowledge. Moreover, as suggested by the analysis of knowledge interfaces in this paper, people at the grassroots may provide a good starting point for engineering such technology interfaces, since they are already combining elements of different knowledge systems in their daily work.

Finally, communication is key to the durability of indigenous knowledge systems. As shown here, in very heterogeneous communities, communication between groups may be too limited to facilitate the passing of more detailed conceptual and technical knowledge. Establishing more platforms for information exchange may thus be important tasks for small grassroots organisations that are sensitive to the cultural parameters of communication. SRISTI's farmer-to-farmer newsletter, the *Honey Bee*, which publishes farmers' ideas and promulgates them to other farmers and extension agents, is one example of this approach. Other avenues may also be available. In the present study area and throughout India, state Forest Departments are currently making concerted efforts to reach out to local communities in an initiative known generally as Joint Forest Management. This initiative involves the creation of village committees which usually consist of one member from every village household and some Forest Department staff. In committee meetings, villagers and foresters are supposed to decide collaboratively on forest regulations and management options for prescribed forest areas close to the villages. In principle, therefore, the Joint Forest Management programme seems quintessentially aimed at bridging gaps between villagers' knowledge systems and official forest science. Provided that forest officers receive proper training in how to foster open, two-way communication across status gaps (see various selections in Poffenberger and McGean 1996) this programme might provide a new arena for interactions of knowledge systems and valuing of local knowledge. In addition, committee meetings might provide new opportunities for villagers who do not always associate with each other to share knowledge about trees and forests. The catalytic power of these meetings was illustrated in one small tribal village in this study. After being introduced to the concept of Forest Protection Committees, villagers became motivated to hold frequent, regular meetings even without forest officers present and to collect a village fund to carry out new projects such as the building of a village temple and the hiring of a school teacher. If conducted in a careful and sensitive manner, therefore, Joint Forest Management holds promise for improving communication and adding value to local knowledge.

Ultimately, people should be able to decide for themselves how their resource management systems are to evolve and which types of knowledge are most useful to them, with 'utility' defined in their own terms. Unfortunately, due to larger-scale processes at work, people at the grassroots level do not always have an optimal degree of control over such matters, nor do they always have the vantage point from which to see how their choices and actions may interact with larger-scale systems and how such interactions will affect them in the long run. As described cogently by Hill (1995: 91),

> Interaction between indigenous and global meanings lies within slowly changing everyday experience, not in a sudden tilt in the world that evokes a clear picture of what is to follow. People simply take advantage of what lies in front of them in a seemingly unproblematic way and *live* the experience, and seek to make sense of the changed meanings with which they are confronted. The *weakness* lies in the fact that the people cannot see what is emerging in the shadows.

While scholars and grassroots organisations by no means have all the answers — nor can they see directly into the future — they may be best situated to point out some of the possible pitfalls hidden in the shadows and to assist people to navigate a safer, more sustainable path through emerging global realities.

Acknowledgements

The author wishes to thank the many village people who kindly gave of their time and expertise to make this study possible.

Fieldwork for this study was supported by the U.S. National Science Foundation under Grant Number SBR-9508550 and the East-West Center. Any opinions, findings, and conclusions or recommendations expressed in this material are those of the author and do not necessarily reflect the views of the National Science Foundation or the East-West Center.

References

Allen, T.F.H. and T.B. Starr, 1982, *Hierarchy: Perspectives for Ecological Complexity*, Chicago: University of Chicago Press.

Bebbington, A., 1996, "Movements, modernizations, and markets: indigenous organisations and agrarian strategies in Ecuador" in R. Peet and M. Watts (eds.), *Liberation Ecologies: Environment, Development, Social Movements*, London: Routledge. pp. 86-109.

Beteille, A., 1996, "Varna and jati", *Sociological Bulletin*, 45(1): pp. 15-27.

Brodt, S., 1998, Learning From the Land: Local Knowledge Systems of Tree Management in Central India, Unpublished Doctoral Dissertation. Department of Geography, University of Hawaii, Manoa, Honolulu.

Brush, S.B., 1992, "Ethnoecology, biodiversity, and modernization in Andean potato agriculture", *Journal of Ethnobiology*, 12(2): pp. 161-85.

Brush, S.B. and D. Stabinsky, 1996, *Valuing Local Knowledge: Indigenous People and Intellectual Property Rights*, Washington, D.C.: Island Press.

Chambers, R., 1983, *Rural Development: Putting the Last First*, Essex, England: Longman Scientific and Technical.

Clarke, W.C., 1990, "Learning from the past: traditional knowledge and sustainable development", *The Contemporary Pacific*, 2(2): pp. 233-53.

Daniggelis, E., 1997, The 'Jangal's' Hidden Wealth: A Survival Strategy by Foraging Farmers in the Upper Arun Valley, Eastern Nepal. Unpublished Dissertation, Department of Anthropology, University of Hawaii, Manoa, Honolulu.

Dash, V.B. and A.M.M. Junius, 1983, *A Hand Book of Ayurveda*, New Delhi: Concept Publishing Co.

den Biggelaar, C., 1996, Farmer Experimentation and Innovation: A Case Study of Knowledge Generation Processes in Agroforestry Systems in *Rwanda*, Rome: Food and Agriculture Organization of the United Nations.

Dove, M., 1992, "Foresters' beliefs about farmers: a priority for social science research in social forestry", *Agroforestry Systems*, 17: pp. 13-41.

Eisemon, T.O., 1989, "Schooling, cognition, and creating capacity for technological innovation in Africa", *International Review of Education*, 35(3): pp. 329-48.

Flavier, J.M., A. de Jesus, and C. S. Navarro, 1995, "The regional program for the promotion of indigenous knowledge in Asia (REPPIKA)" in D.M. Warren, L.J. Slikkerveer, and D. Brokensha (eds.), *The Cultural Dimensions of Development: Indigenous Knowledge Systems*, London: Intermediate Technology Publications. pp. 479-87.

Friedman, J., 1989, "Culture, identity, and world process", *Review*, 12(1): pp. 51-69.

Gadgil, M. and R. Guha, 1992, *This Fissured Land: An Ecological History of India*, Berkeley: University of California Press.

Gadgil, M. and R. Guha, 1995, *Ecology and Equity: The Use and Abuse of Nature in Contemporary India*, London: Routledge.

Gogoi, D.K., 1990, *Agricultural Communication Networks: A Village Level Analysis of Punjab*, New Delhi: Radiant Publishers.

Gupta, A., 1996a, "Rewarding creativity: a case for registration of small innovations", *Honey Bee*, 7(3): pp. 1-2.

Gupta, A., 1996b, "The Honey Bee network: voices from grassroots innovators", *Cultural Survival Quarterly*, Spring: pp. 57-60.

Gupta, A., 1997, "Creativity, innovation, entrepreneurship and networking at grassroots level" in B. Singh and N. Mahanti (eds.), *Intellectual Property Rights and the Tribals*, New Delhi: Inter-India Publications. pp. 67-80.

Gupta, A., 1998, "India innovates: an open letter to parliamentarians", *Honey Bee*, 9(1): pp. 1-2.

Hill, S., 1995, "Globalization or indigenization: new alignments between knowledge and culture", *Knowledge and Policy*, 8(2): pp. 88-112.

Holling, C.S., 1986, "The resilience of terrestrial ecosystems: local surprise and global change" in W.C. Clark and R.E. Munn (eds.), *Sustainable Development of the Biosphere*, Cambridge: Cambridge University Press.

Horton, R., 1982, "Tradition and modernity revisited", in M. Hollis and S. Lukes (eds.), *Rationality and Relativism*, Cambridge, MA: The MIT Press. pp. 201-60.

Howes, D. (ed.), 1996, *Cross-Cultural Consumption: Global Markets, Local Realities*, London: Routledge.

Howes, M. and R. Chambers, 1980, "Indigenous technical knowledge: analysis, implications, and issues" in D.W. Brokensha and D.M. Warren (eds.), *Indigenous Knowledge Systems and Development*, Lanham, Maryland: University Press of America. pp. 329-40.

Ingold, T., 1992, "Culture and the perception of the environment", in E. Croll and D. Parkin (eds.), *Bush Base: Forest Farm: Culture, Environment and Development*, London: Routledge. pp. 39-56.

Jolly, J., 1977, *Indian Medicine*, New Delhi: Munshiram Manoharlal Publishers Pvt. Ltd.

Kalland, A., 1994, "Indigenous knowledge — local knowledge: prospects and limitations", in B. Hansen (ed.), Arctic Environment. A Report on the Seminar on Integrating Indigenous Peoples' Knowledge, September 20-23, 1994. Copenhagen: Ministry of the Environment/ The Home Rule of Greenland. pp. 150-67.

Kangayappan, K., 1992, "The caste system: social stratification in India", *Asian Thought and Society*, 17(49): pp. 42-50.

Mahanti, N., 1997, "Theme paper", in B. Singh and N. Mahanti (eds.), *Intellectual Property Rights and the Tribals*, New Delhi: Inter-India Publications. pp. 29-35.

May, R.M., 1973, *Stability and Complexity in Model Ecosystems*, Princeton, New Jersey: Princeton University Press.

Poffenberger, M. and B. McGean (eds.), 1996, *Village Voices, Forest Choices: Joint Forest Management in India*, Delhi: Oxford University Press.

Rani, S. and A. Malaviya, 1991, *Communication and Rural Women*, Delhi: Manak Publications.

Richards, P., 1985, *Indigenous Agricultural Revolution: Ecology and Food Production in West Africa*, London: Hutchinson.

Rogers, E.M., 1995, *The Diffusion of Innovations* (Fourth Ed.), New York: Free Press.

Shiva, V., 1997, *Biopiracy: The Plunder of Nature and Knowledge*, Boston: South End Press.

Singh, B. and N. Mahanti (eds), 1997, *Intellectual Property Rights and the Tribals*, New Delhi: Inter-India Publications.

Sinha, D., 1995, "Caste as an economic institution in India: an interdisciplinary study", *Asian Economies*, 24(1): pp. 27-45.

Thrupp, L.A., 1989, "Legitimizing local knowledge: from displacement to empowerment for Third World people", *Agriculture and Human Values: Summer*, 1989, pp. 13-24.

5

Eco-Diversity and Indigenous Knowledge on Forests

Nityananda Patnaik

The legacy of Indian forest culture

THE paper refers to the ideas of both rural and tribal people with regard to the forests in their regions. Forest is an area set aside for the production of wood and other forest produce. It is maintained and protected to meet the various needs of man and indeed also for all other living organisms.

The pastoral Aryans who penetrated into India around 2000 BC lived in consonance with the thick extensive forests teeming with wild animals. Being worshippers of nature, they preferred sylvan surroundings as their abode. They cleared the forest and setup their shelter. It was in such settings that the *Veda*s and the *Upanishad*s were composed by them. Many cultural and educational centres patronised by the Aryans came up at various places in the inspiring landscape and many saints started preaching the precepts of conduct for man to live a righteous life. There is evidence that the Aryans had encounters with many tribal communities who had already lived in those areas prior to the Aryan inroads into this country. Being autochthonous and the earliest settlers in the forest areas for thousands of years, they were enjoying considerable freedom to use the wood, to exploit the minor forest produce and to hunt animals and they had an ineradicable conviction that the forest was theirs.

The great epic *Ramayana* has given reference to 'Namisharanya', 'Chitrakuta' and 'Dandakaranya' which were very thick dense forests abounding with wild animals. But due to the increase in population and the demands of commerce, the onslaught on the vegetation started though in the beginning slowly but steadily. The first reference to the destruction of forest was the burning of 'Khandava-vana' mentioned in the *Mahabharata*.

Being very much concerned about the destruction of the forest many of our wise ancestors took steps to stop this process of extinction. One of them was Kautilya, who during his time raised forests of various types such as forests for 'Soma' plantations, forests of delicious fruit trees, forests with trees and bushes for fuel and fodder, forests with wild animals, forests for elephants. In some places, particular trees were planted in large numbers to satisfy the food requirements of the animals.

During the time of Emperor Ashoka, the cultivation of exotic medicinal plants was taken up. The rock edict of Ashoka provides information regarding the plantation of medicinal plants in areas where these were not present. The Moghuls and Marathas and the Gonds were great lovers of trees. During their periods, mango groves and many other useful trees were planted in various places and on road sides. In these earlier days, our ancestors considered cutting of trees a sin and planting and protecting them an act of piety. The avenue plantation of very large trees on either side of the main arteries of travel and the large mango groves we find today in ex-Princely States were the works of the rulers and kings of bygone days. Since time immemorial, many useful trees have been considered sacred. Such trees like banyan (*Ficus benghalensis*), *pipal* (*Ficus religiosa*), *bel* (*Aegle marmelos*), etc. have been saved from extinction. Some other species such as the *kadama* (*Anthocephalus cadamba*), the *ashoka* (*Saraka indica*), etc. have, however, disappeared.

Forests of Orissa

Orissa is situated within the subtropics and therefore all the forests are of subtropical forest types. In most of the forests *sal*

(*Shorea robusta*) constitutes the dominant species. However, in some areas, *sal* does not occur at all or occurs scarcely. Orissa is the southernmost region of *sal* forests in India. Being located midway between the north and the south, the state enjoys the unique privilege of having both southern and northern subtropical forest types. Orissa is endowed with forests covering a total area of 66,400 sq km. Regarding the extent of land under forest the figure varies from source to source. The official record shows that 42.65 per cent of the total geographical area is under forest of one kind or the other. The bulk of more valuable forest is located in the hilly and undulating mountainous tracts of the northern plateau and in the Eastern Ghats, where almost all tribal communities of Orissa live in large concentrations. In other words, the forests and the tribals coexist in different ecological niches of the state. Table1 gives the districtwise extent of land under forest and their tribal population.

The ethnographic scene

There are as many as 450 different tribal communities inhabiting different parts of India. The diversity of these ethnic groups in terms of race, language and culture is so great that a generalised presentation of the whole fabric of tribal life and culture is impracticable. It is for this reason that I have concerned myself with a few cases from the state of Orissa, each dealing with a specific tribal society and with particular problems relating to the symbiotic relationship between them and the forest and their indigenous knowledge by which they eke out their living in their respective habitats. In the north of the state, the blocks of mountainous country in Keonjhar, Bonai, Pallahara and Similipal hills and in the south the immense tracts of mountainous country comprising the Ganjam Agency tracts, the whole of Koraput district, major parts of Phulbani district and the southern Dongar lands of Kalahandi district are the areas of the largest tribal concentration in the state. Some of the tribes inhabiting the northern highlands are the Bhuinya, the Juang and the Kondh of the Bamara-Redhakhol border.

Table 1: Forest land and tribal population (Source: Census of India, 1981)

Name of the district	*Geographical area (in ha)*	*Forest land (in ha)*	*Tribal population*
1. Sambalpur	1749 (100%)	701 (40.08%)	(27.4%)
*2. Sundargarh	979 (100%)	549 (56.08%)	(50.7%)
*3. Keonjhar	831 (100%)	407 (48.98%)	(44.5%)
*4. Mayurbhanj	1040 (100%)	468 (45%)	(57.8%)
5. Balesore	647 (100%)	40 (6.18%)	(7.0%)
6. Cuttack	1119 (100%)	140 (12.51%)	(3.2%)
x7. Dhenkanal	1092 100%)	454 (41.58%)	(12.1%)
x8. Phulbani	1104 (100%)	829 (74.09%)	(37.3%)
9. Balangir	883 (100%)	213 (24.12%)	(18.5%)
x10. Kalahandi	1158 (100%)	527 (45.51%)	(30.9%)
*11. Koraput	2702 (100%)	1400 (51.81%)	(54.3%)
x12. Ganjam	1220 (100%)	569 (46.64%)	(9.3%)
13. Puri	1046 (100%)	343 (32.79%)	(3.5%)

* Areas having extensive forests with large tribal concentration.

x Areas having extensive thick forests with large tribal concentration only in hilly regions of the Great Line of the Eastern Ghats.

The tribes inhabiting the mountainous regions in southern Orissa are the Bondo and the Didayi of the Bondo Hills, the Dongria and the Kuttia Kondh of Niyamgiri and the Belghar Hills, the Lanjia Saora and Malia Kondh of Mahendragiri and the Khondmal Hills. The Kuttia Kondh and other divisions of the Kondh tribe are inhabiting the hilly areas of Thuamul Rampur and Kashipur regions. All these mountains form the northern extremity of the Eastern Ghats and are covered with very thick forests.

THE HUNTING AND FOOD GATHERING TRIBES

The Birhor (literally meaning: man of the forest and hills) are a semi-nomadic tribe with a total population of about 1,349 persons distributed mainly in the northern districts of the state. This tribe is locally known as Mankidi, Mankirdia, Mankria or Mankarkhia Kolha meaning monkey eaters.

They live in small leaf huts called Kumbha set-up at the outskirts of the forest-clad hills. Generally 10 to 20 families related by kinship ties live together in a place in their respective small conical leaf huts. This temporary camp is known as Tanda. The Tanda, representing the band life of the Birhor, is located invariably in places where water sources, thick forests, peasant villages and weekly markets are located nearby.

The whole rubric of the Birhor subsistence base involves on the one hand the exploitation of forest resources and on the other the economic interaction with the larger peasant society. From the forests they get two dozen varieties of animal foods and many types of plant foods. They also collect larvae of different ants and other insects, honey, roots and tubers, various fruits, different algae, mushrooms and bamboo shoots. The green leaves of such types as *munga* (*Moringa olefera*), *pankar* (*Ficus cordifolia*), *koenar* (*Bauhinia purpurea*), *amti* (*Antidesma diandrum*), *kehla* (*Grewia tilaefolia*), etc., form their essential food items.

The Birhor are excellent monkey catchers. Mainly, *Macacus rhesus* and *Presbytes entellus* varieties of monkeys are caught with special nets prepared by them. They eat monkey flesh and trade with monkey skin.

Another speciality of the Birhor is rope making. The fibres of Siali creepers (*Bauhinia vahlii*) which they collect from the forest are used to make ropes and rope-made products. They have specialised on this manual work and it is their primary source of income.

The forest also provides them with certain medicinal herbs such as Mahajatta (used for curing fever), roots of biar (used for curing stomach ache), roots of banna (used for curing dysentery), Parhi (used for curing fever), and Rohain (used for curing headache and fever). They sell these jungle herbs as medicines and also work as medicine men for the neighbouring people.

As pointed out above, the Birhor do not live permanently in a particular place. As the season changes, they shift their camp to some other place in another district of the state. They are knowledgeable about what forest resources are available in which place in the different seasons and accordingly they move to such places to make use of them.

SHIFTING CULTIVATION

Shifting cultivation is a mode of agriculture in which impermanent clearings on hill-slopes are cropped for a certain period and then they are left fallow. There are many areas in Orissa, as indeed in other parts of India and sub- and tropical countries, where the terrain is only suitable for shifting cultivation. This type of tillage is also known as slash-and-burn or swidden cultivation. It is called Podu among the Khond, Bagada among the Saora, Kaman or Bringa among the Juang and Paudi Bhuinya in Orissa. It is commonly known as Jhum in north-east India.

The land under shifting cultivation belongs to the village and it is the village elders who decide in a village meeting which site they should cultivate and how to distribute the patches among themselves according to their requirements. Almost all activities connected with shifting cultivation are performed by communal labour.

It is commonly believed that many years back man observed that seeds of certain plants had fallen on the ground and sprouted and given rise to plants which bore edible fruits. To obtain these easily he cleared a patch of forest near his abode, burnt the debris and grew such plants. Later he discovered that the forest soil mixed with ashes gave bumper crops. He found out that repeated cropping in one patch diminished the crop yield and therefore he took up another patch of forest and repeated the same process. This land use system involves the cutting and clearing of the vegetation on a hill-slope, burning the debris and dried slash before the onset of the monsoon, mixing the ashes with the soil by means of a hand hoe, raising a mixture of crops for a couple of years and keeping it fallow thereafter for a period varying from four to six years and eventually returning to the same plot for another cropping phase. A few decades ago the recuperative cycle was for a period of 20 years, which has been shortened to four to six years due to an increase in population and a shortage of land.

Shifting cultivation begins with forest-clearing. Barring a few communities, who cut the trees down to the ground, most of the communities cut the trees high above the ground leaving the stumps of about 2 ft in height. The coppicing of such stumps produces green foliage which people use for food. The tribes of Orissa work the soil by means of a hoe tipped with an iron point. In South Orissa ploughing is done parallel to the contour because people have observed that the soil runs off the surface and is washed down by the rains when the hill-slope is ploughed from top to bottom. In fact the higher slopes of the hills up to the tops in South Orissa are composed of Khondalite rock yielding a friable soil particularly susceptible to erosion. In North Orissa the rocks yield red-clay-loam, as people say, which erodes slowly and stands up to the onslaught of axe-cultivation.

A common practice followed in shifting agriculture is the simultaneous sowing of multiple seeds mixed together and the sequential harvesting of crops beginning in September and ending in December. Some communities grow more than 50 varieties of crops, most of which are minor varieties of millet, in a single swidden strip.

Some advantages of the practice of mixed cropping, according to the shifting cultivators' opinion, are that the difficulty of food storage is solved and food shortage is evaded for a longer period of time. From an agronomic point of view not shared by tribal people the crop mixture provides a good crop cover preventing nutrient loss of the soil through run off and leaching of water and helps in optimising resource use of a single plot of land. The next crop gets more space after the harvest of the earlier ones and it ensures the recycling of biomass and nutrients through crop and weed residues and enriches the soil through weed mulching.

The economic efficiency of shifting agriculture is related to the length of the recuperative cycle. The yield declines with the shortening of the shifting cycle. The people's experience is that the amount of labour involved in various operations outweighs the yield in the sequential agricultural cycle and it is the labour input and not the soil erosion and diminishing yield which compels them to shift to another plot.

One point which the swidden cultivators stressed with respect to mixed cropping is that it is a way of pest control. By doing so, the farmers minimise the risk of a bad harvest. Distance in space and time between co-specific plants prevents danger to the crops from herbivores. Moreover the mixed cropping enables the swidden cultivators to use some grains as food and some for clearing debts.

TERRACED CULTIVATION

The Lanjia Saora, one of the primitive tribes of Orissa, are found in Ganjam and Koraput districts. In addition to shifting cultivation they grow paddy on terraces built on hill-slopes right from the foot-hills to the hill-top. The construction of these terraces exhibits the wonderful patience and skill of those who built them. The Lanjia Saora are very good cultivators skilled in revetment, terracing and water management.

The explanations which they give in favour of terraced cultivation are that the hill-slopes are composed of Khondalite rocks which are susceptible to erosion and therefore the soil formed by it is pealed off from the surface rendering it unsuitable for carrying on shifting cultivation. In such a situation terracing is thought to be the best method for checking soil erosion. Terraced fields with permanent irrigation are laborious, but provide the peasants with higher yields.

In another region, which is similar in terrain and climatic conditions to that where paddy cultivation is practised on terraces, the same tribe has taken up orange growing on terraces. Here bench terraces instead of table terraces have been built for orange plantations. It was a Lanjia Saora man who raised seedlings in his home garden and planted them in a terrace uphill in a small number. He was successful in his experiment and continued it on a larger scale in subsequent years. Gradually the whole area and now the whole Mahendragiri hill ranges are covered with orange orchards replacing shifting cultivation.

It goes without saying that the land utilisation pattern in a particular area must be in conformity with climatic conditions. The knowledge and wisdom which man possesses help him to be innovative and to achieve a kind of cropping pattern favoured by a particular climate. This is what has happened in this case. The high altitude, good drainage and longer duration of the low temperature period favour the growth of citrus plants. These conditions prevail in the Mahendragiri region and have helped the farmer in his efforts.

There is a similar example of an indigenous vision of the Dongria Khond, a Kui-speaking section of the large Khond tribe inhabiting the Niyamgiri hill ranges of the Koraput and Kalahandi districts. The Dongria Khond are famous horticulturists. The whole of the Niyamgiri hill ranges, which is their settlement area, are covered with plantations of mango, jackfruit, pineapple, orange and banana. The Dongria Khond are very systematic planters of fruit trees and at the foot-hills they grow turmeric. The land above the turmeric patch is used for growing orange plants. Generally, orange trees are grown in places close to a river. Along with the orange plants, mango and jackfruit trees are also grown in the orange orchards. Higher upwards pineapple is planted and jackfruit trees are grown here and there in the pineapple plantation to provide them with shade. Banana plants are grown in the outskirts of the pineapple plantation. A field hut is set up at the hill-top for the farmer to take rest and watch the plantations. These plantations of different fruit trees in different altitudes, in different strips situated horizontally from the foot-hill almost to the hill-top resemble a modern agro-ecosystem called Sloping Agriculture Land Technology (SALT). This agro-horticulture system was achieved by the indigenous vision of the Dongria Khond.

Forest and tribe

The tribal people who have been living for thousands of years in the forest have a very strong conviction that the forest is theirs. They were enjoying considerable freedom to use wood, exploit minor forest produce and hunt animals. This freedom has been curtailed with the recent forest policies aimed primarily at protection and conservation of the forests in the national interest. Nevertheless the tribal people continue to feel that they have now the same rights in making use of the forest to meet their needs as they used to have before.

Being associated with forest since time beyond memory their cultures are forest-oriented. Many of their folk-tales are about trees and animals of the forest. Their clan organisation and marriage rules are intimately associated with vegetation and animals. The inspiring landscape of the sylvan surrounding in which they live has made them creative, artistic and knowledgeable about various cosmic phenomena. These talents are reflected in their songs, dances, arts and crafts, in their work and thought and in various other aspects of their life.

Tribal people observe many taboos when they use trees. To cut mango, *mahua* (*Madhuca longifolia*), tamarind and many other fruit-bearing species is taboo among tribal communities. This taboo has taken its origin from certain deities who are said to be great lovers of trees. A folk-tale related to the origin of trees collected from the Lanjia Saora illustrates this point.

After the first earth had sunk beneath the waters and the new earth was made, Kittung, one of the deities who is the great lover of trees planted the forest with fruit-bearing trees. These were the only trees he made. At that time there were few people and no grain. Everybody lived on fruits. But when the population increased and men made fields and hill-clearings and sowed them with grains, they needed firewood to cook with and they cut the fruit-trees and burnt the wood. In this way the forest was destroyed. When Kittung saw this he was very sad. He went out to find seeds to make new ones. At last Kittung got the seeds of new trees from one of his tutelaries. He sowed the seeds and when the forest had grown up again he said: "Don't cut fruit trees till they are dry and useless. Cut the new trees I have given for your firewood and house."[1]

A few examples of trees are given below to show how they are used by the Saora:

1. *Aegle marmelos*, Corr. (Saora term: Kulpatneban): This is the *bel* tree. Its leaves are used in divination.

2. *Artocarpus integrifolia*, Linn. (Saora term: Padasaneban): This is the jackfruit tree. Its fruits are a very popular food.

3. *Bassia latifolia*, Roxb. (Saora term: Abaneban): This tree provides food, oil and liquor. It is taboo to cut it.

4. *Bauhinia vahlii*, W.et A.: The leaves of this creeper are used for making leaf plates.

5. *Bombax malabaricum* (Saora term: Kuikuineb): The cotton tree. In a cholera epidemic the wood is used to make a little chariot in which Mardisum (cholera deity) is escorted out of the village.

---

1. Elwin, Verrier, 1955: *The Religion of an Indian Tribe*, Oxford University Press, pp. 204-5.

6. *Butea frondosa*, Roxb. (The flame of the forest): The flowers are offered to the ancestors.
7. *Diospyros melanoxylon*, Roxb.: The Ebony tree is locally believed to have grown from the grave of a girl who had been bitten by a snake. In some villages, a pole of Ebony is used to protect tobacco or chilli gardens.
8. *Eugenia jambolana*, Lam.: The black plum. A branch of this tree is used to sprinkle medicated water to drive away any kind of cattle disease.
9. *Ficus benghalensis*: The banyan is one of the sacred trees. Its leaves are used for ceremonial purposes. Offerings are given at the harvest festival to it.
10. *Ficus religiosa*: The *pipal* is a sacred tree.
11. *Holarrhena antidysenterica*, Wall: Offerings are made on the leaves of this tree.
12. *Mangifera indica*: The mango tree. Its fruits are eaten and the leaves are used in ritual.
13. *Malia azadirachta*, Linn.: The bitter leaves are used in the feast when a person has died.
14. *Musa sapientum*: The broad leaves of the plantain are used in many sacrificial events.
15. *Phyllanthus emblica*, Linn.: It is used in rituals.
16. *Pongamia glabra*, Vent.:The wood of this tree is commonly used for funeral pyres.
17. *Semecarpus anacardium*, Linn.: The leaves are considered efficacious against soreness. A suitable offering to the tree may be made for any suffering from sore eyes.
19. *Shorea robusta*, Gaertn.: The images and birds on a shrine are made of this wood. The leaves are used in any sacrifice.

20. *Tamarindus indica*, Linn.: The fruits are used as pickles.
21. *Trewia nudiflora*, Linn.: Amulets are made of the bark of this tree.

SACRED GROVES

Sacred groves are one of the finest instances of traditional conservation practices and they also serve as centres of cultural and religious life. One example of sacred groves is drawn from the Santal community. Two characteristic features of a Santal village are the Jaherthan (sacred grove) and the Manjhithan (abode of ancestors of village headmen). The Jaherthan or sacred grove is regarded as the abode of the principal Santal deities and is situated at the end of the village. It is a large patch of forest containing *sal*, *mahua* and other types of trees. No one is allowed to cut any tree from this grove. Madhav Gadgil and Subash Chandran[2] have mentioned in their landscape ecological studies in Uttar Kannada district that sacred groves are one of the finest instances of traditional conservation practices. When community management of natural resources prevailed in the pre-British period, there existed a mosaic of pristine sacred groves. Due to the population explosion and over-exploitation of forest resources for commercial purposes, numerous sacred groves were included in forest-working plans for selective felling. Planting of coffee and tea and the plantation of Eucalyptus have wiped out a number of sacred groves.

MEDICINAL HERBS

A study of tribal health and medicine leads one to the study of medicinal plants which the tribal people use as curative and preventive measures of treatment. This study shows that the tribals have a very deep knowledge of various plants and it

---

2. Gadgil, M., & Chandran, Subash, 1989: *Environmental impact of forest-based industries on the evergreen forests of Uttara Kannada district: A case study*, Final report, submitted to Department of Ecology and Environment, Government of Karnataka.

explains how they use such plants to cure diseases. Rev. P.O. Bodding who came to India as a missionary at the beginning of this century has mentioned in his work on Santal medicine thousands of plants and plant-products which were used by the Santal for the treatment of diseases.[3] Works of this nature show how deeply the Santal are associated with the natural world and how deep their knowledge is in making use of the plant products for their health and well-being.

DIET

An inventory of food and drinks of the tribal people reveals that the major portion of their diet is composed of forest products. According to their needs and local resources, there are wide variations in the food habits of the tribal people living under different climatic conditions. Those who have taken up settled cultivation depend primarily on rice, cereals, millets and vegetables. Some tribal communities living in the forest depend largely on roots and tubers, fruits and berries, edible flowers, leaves and honey, insect larvae, wild birds and rats. Other important sources of food supply for the forest dwellers are wild as well as their domestic animals. Thus among tribal communities there are vegetarian tribes and non-vegetarian tribes. In the case of the Toda of the Nilgiri hills, the diet is largely of the lacto-vegetarian type. One of the noted purely vegetarian tribe are the Hunza of Karakoram who subsist on wheat, barley, beans, grams, pulses, apricots, milk, ghee and whey.

The food habits of different tribal groups reflect differences in their cultural patterns. The food getting activities, different types of fermented beverages such as *handia* (rice beer), *mahuli* (liquor distilled from flowers of *Bassia latifolia*), etc., tapping of toddy and juice from sagopalm for the observance of rituals and ceremonies are only a few instances to show how the tribal cultures are closely interlinked with their nutriment needs. Those tribal groups living in the remotest corners as forest dwellers

3. Bodding, Rev. P.O., 1940: *Memories of the Royal Asiatic Society of Bengal.*

and those at a hunting and food gathering stage depend entirely on forest products for their subsistence. The tribals who practice shifting cultivation partly depend on their agricultural produce and partly on forest produce for their food. The food of the tribals who have taken up regular wet-cultivation in the plains and who work in industrial areas is mainly composed of agricultural products.

A list of different varieties of flowers, fruits, leaves, vegetables, roots and tubers consumed by the tribal people in different months is given in the annex.

INDIGENOUS KNOWLEDGE OF SOME COSMIC PHENOMENA

The tribal people and indeed perhaps all human beings look upon earth (Skt.: *bhumi*) with great reverence. Some of them even treat themselves as children of the soil and worship the earth thinking it to be their mother (Mother Earth). Most of the tribal communities live in villages and utilise the resources available within the boundary of their respective villages. They know a lot about the types of soil and the crops which can be grown at best. What crops can be grown in the hill-slopes, at the foot-hills, in uplands and backyards and in valleys is determined by the texture of the soil. To retain moisture the stones lying on the surface in a patch of land are not removed and orange trees, for instance, are grown in such rocky soil.

Various environmental phenomena indicate the approach of the rainy season. The tribal people can predict the approach of the rains, the volume of rain and the duration of the rainy season from these indications. The profuse fruiting and immediate ripening of fruits of *Syzygium cumini* and profuse flowering of *Lagerstroemia microcarpa* are sure signs of the coming rain. Mass movements of ants indicate that the rains are around the corner. A kind of spider weaving its web is indicative of the end of the rainy season. The tribal people have observed that the mango trees and other trees standing on either side of the hill stream have green leaves. By observing these trees one can locate the sources of water in the forest. The presence of ground water can be predicted by observing the bright green foliage of *Ficus benghalensis*. The wet soil of an anthill indicates that water is available at a short depth.

The tribal people are particularly selective in the use of various types of wood. Certain species are good for fuel and for making charcoal, some others for making ploughs and other agricultural implements, some for making doors and windows and some for making furniture. The tribals are very much concerned about the processes and consequences of deforestation. The orange growers of the Mahendragiri hills are worried about the climate getting hotter due to the depletion of forests which affects their orange fruits. With the rise in temperature the yield is diminishing and the fruits are getting smaller and smaller in size.

An observation of the way of life of the tribal people shows that without exception, man in a tribal society lives as part and parcel of this ecosystem. He modifies it for his survival and at the same time tries not to disturb its balance. He does not live in isolation in the microcosm of his ecosystem but is linked with other ecosystems, all taken together forming a landscape. He clears the forest for cultivation and conserves it by following many taboos and by personifying nature and establishing a social relationship with it to meet his needs. He works in such a manner that biodiversity is maintained, heterogeneity is increased and the forest produce made available is plenty.

Another side of this interlinkage is his relationship with celestial bodies. The sun is his benevolent supreme deity and it is worshipped on all festive occasions. He plants in the sunlight to improve the growth of his crops. He understands fully the utility of sunlight in the growth of the forest. Different stars give guidance to him in his agricultural operations. The winds indicate to him the seasonal changes. The integration of man-nature-supernatural, the components of cosmos, needs a much closer and deeper study for a better understanding of indigenous knowledge about the resource management system of tribal people.

Conclusion

In a nutshell the indigenous knowledge of tribal people on forests is summarised below.

Tribal people live in harmony with their environment and their culture is incorporating the spiritual and the material, the living and the non-living into one integral whole. Being part of living nature the environment is not exploitable. This holistic concept is evident and expressed in their myths and legends and in their wall paintings and decorations. The conservation of many plants and animals in the forest environment is a part of the tribal culture. The taboos and customs enforce protection and conservation of forests teeming with a wide diversity of species which are not destroyed. Many songs composed by the tribal people are full of respect for plants and animals. The trees and hills are depicted in the icons and wall paintings.

The tribal people are full of knowledge about various uses of hundreds of plants. They know which plants have nutrients and medicinal values and which plants are good for fuel and for agricultural equipment. Some tribal people control pests by planting branches of trees in their paddy fields for birds to perch on these trees. Some plants are used to fight pests. Certain diseases of paddy plants are controlled by putting the leaves of *khair* (*Acacia catechu*) in the water channel which flows into the paddy fields.

Several cattle diseases are treated with herbal remedies such as the leaves of *sitaphal* (*Annona squamosa*) or *karela* (*Mordica chrantia*) and the seeds of *palasa* (*Butea monosperma*).

Tribal people roam in the forest for roots to live on in times of scarcity. The *mahua* flowers (*Madhuca indica*) which they eat provide sufficient carbohydrate to them and the berries of *karandi* (*Carisa congesta*) are eaten raw or ripe. They are so familiar with plants and trees of the forest that they can easily identify which plants and plant products are edible and which are harmful. Animals and birds are hunted by them in their forest habitat. The nets and the traps which are used for catching fish and birds are made of fibres which they get from certain tree species found in the forest. Sometimes traps and nets are set in trees and leaves, fruits, roots or barks of certain plants are thrown in the streams to stupefy fishes.

The houses are constructed with several types of timber collected from the forest. Many tribal people use a kind of grass grown in the forest for thatching their houses. Reeds and twigs of bamboo are used as materials for making walls. Some tribal communities use both bamboo and planks to make the walls. The wattle walls are plastered with mud mixed with cow-dung.

The tribal people collect fuel-wood from the forest for their daily needs. The women are very particular about the type of firewood, preferring that which burns well even when fresh. They avoid as far as possible those woods which emit smoke. Recent developments in the sphere of forestry which give more stress on monocultures of species have completely overlooked the concern for women's health.

Generally the tribal people do not possess modern torches for lighting. While going from one village to another in the dark, they use certain woods as a torch. There are many plants in the forest which are used for getting fibres used for making cords and ropes. The Siali creeper (*Bauhinia vahlii*) is such a plant with strong fibres with which small baskets and ropes are made. Tasar silk worms are reared in the *sal* trees (*Shorea robusta*) grown in forests. The medicine men as a part of training help their trainees to get acquainted with the medicinal herbs while roaming in the forest during fieldwork. The tribal people are habituated to the collection of various minor forest produce such as resins, lac, seeds, honey, bamboo, reeds, grasses, *mahua* seeds, *sal* seeds, marking nuts, myrobalan (*Terminalia chebula*, *Terminalia bellerica*, *Emblica officinalis*), hillbrooms, wax, and various medicinal herbs, etc. from the forest either for their own use or for sale. Among the oilseeds collected from the forest are those of *mahua*, Karang (*Pongamia pinnata*), Kusum (*Schleichera oleosa*), Polang (*Calophyllum inophyllum*). Mats are made out of palm leaves. The Sago palm (*Caryota urens*) and ordinary palm trees are tapped for the sap which they drink as a regular daily habit. The palm wine is considered a generous and abundant gift of Mother Earth.

There is an age-old symbiotic relationship between the tribal people and the forests. This close relationship has made them rich in knowledge about the trees, the properties and utilities of different plants, the seasonal changes in the life-cycle of the plants and the natural phenomena influencing them. This indigenous knowledge about the nature and natural phenomena with particular reference to forests and forest products has been a great help to the tribal people for their very survival and existence against all odds, calamities and mishaps which they encounter in their day-to-day life.

ANNEX

Seasonal calendar of flowers, fruits, leaves, vegetables and roots. (All vernacular terms are in Hindi)

Month in which obtained	*Item*	*Name of article*
1. March-April	Flowers	Gulgul (*Cochlospermum gossypium*), Mahua (*Madhuca indica*).
	Fruits	Piar (*Buchanania latifolia*), Kend (*Dyospyros melanoxylon*), Kanand (*Carissa carandas*), Bel-chenjo (*Aegle marmelos*),Mango (*Mangifera indica*),Kakri, Kharbuja (*Cucumis melo*).
	Leaves	Gendhari (*Amaranthus gangeticus*), Bathua, (*Chenopodium album*), Koinar (*Bauhinia purpurea*)
	Vegetables	Brinjal (*Solanum melongena*),Kareli (*Momordica charantia*) (Bitter gourd), Laoka (*Lagenaria vulgaris*), Semi (Bean)
	Roots	Gethi (*Dioscorea bulbifera*), Khania (*Dioscorea pentaphylla*), Kulu (*Dioscorea Daemona*), Dura (*Dioscorea belophylla*) and other types of tubers.
2. April-May	Flowers	Kachnar (*Bauhinia variegata*), Mahua (*Madhuca indica*), Jirhul (*Lagerstroemia regina macro carpa*)

Month in which obtained	*Item*	*Name of article*
	Fruits	Wild Bel (*Aegle marmelos*), Piar (*Buchanania latifolia*), Kendu (*Dyospyros melanoxylon*), Sarai (fruit of *sal* tree), Pakua (fruit of *pipal* tree), Kharar (*Gardenia turgida*), Jamun (*Eugenia jambolana*), Mango (*Mangifera indica*), Jackfruit (*Artocarpus integrifolia*), Kakri (*Cucumber*), Kharbuja (*Cucumis melo* or water melon), Paper (*Pongamia glabra*).
	Leaves	Chakwar (*Antidesma diandrum*), Koinar (*Bauhinia purpurea*), Gendhari (*Amaranthus gangeticus*).
	Vegetables	Brinjal (*Solanum melongena*), Kareli (Bitter gourd), Laoka (*Lagenaria vulgaris*), Semi (Bean).
	Roots	Gethi (*Dioscorea bulbifera*), Dura (*Dioscorea belophylla*), Khania (*Dioscorea pentaphylla*), Kulu (*Dioscora daemona*)
3. May-June	Fruits	Sarai (fruit of *sal* tree), Kendu (*Dyosphyros melanoxylon*), Piar (*Buchanania latifolia*), Pipul (*Ficus religiosa*), Bar (*Ziziphus jujuba*), Bel (*Aegle marmelos*), Barhar (*Artocarpus lackoocha*), Kat-Jamun (Eugenia heyneana).
	Leaves	Chakwar (*Antidesma diandrum*), Koinar (*Bauhinia purpurea*), Gendhari (*Amaranthus gangetica*), Pipal (*Ficus religiosa*), Pankar (*Ficus cordifolia*).
	Vegetables	Kareli (Bitter gourd), Laoka (*Lagenaria vulgaris*), Kohunra (*Cucurbita*)
	Roots	Khania (*Dioscorea pentaphylla*), Dura (*Dioscorea belophylla*), Kulu (*Dioscorea daemona*), Gethi (*Dioscorea bulbifera*).
4. June-July	Fruits	Kussam, (*Schleichera trijuga*), Dori (fruit of *Madhuca indica*), Bar (*Ziziphus jujuba*), Bel (Aegle *marmelos*), Dates (Khejur), Mango (*Mangifera indica*), Jamun (*Eugenia jambolana*), Kathar (*Artocarpus integrifolia*).
	Leaves	Gendhari (*Amaranthus gangeticus*), Chakwar (*Antidesma diandrum*).

Month in which obtained	*Item*	*Name of article*
	Roots	Khania (*Dioscorea pentaphylla*), Dura (*Dioscorea belophylla*), Gethi (*Dioscorea bulbifera*).
5. July-August	Fruits	Jamun (*Eugenia jambolana*), Sikat (*Breynia rhamonoides*).
	Leaves	Chakwar (*Antidesma diandrum*), Gendhari (*Amaranthus gangeticus*).
	Vegetables	Kareli (bitter gourd).
	Roots	Gethi (*Dioscorea bulbifera*).
6. August-September	Leaves	Amti (*Antidesma diandrum*).
	Vegetables	Kareli (bitter gourd), Cucumber, Nenua (*Luffa aegyptica*), Banskhukri (mushrooms).
	Roots	Gethi (*Dioscorea bulbifera*), Pathal Kohnra (*Pueraria tuberosa*).
7. September-October	Leaves	Palki (*Spinacia oleracea*).
	Vegetables	Brinjals (*Solanum melongena*), tomatoes (*Lycopersicum esculentum*), Aru (Yams), Jhingi (*Luffa acutangula*), Khira (Cucumber) Kareli (bitter gourd).
	Roots	Gethi (*Dioscorea bulbifera*), Dura (*Dioscorea belophylla*), Khania (*Dioscorea pentaphylla*).
8. October-November	Fruits	Bar (*Ziziphus jojoba*), Sarifa (*Anona sqamosa*).
	Leaves	Palki (*Spinacia oleracea*).
	Roots	Khania (*Dioscorea pentaphylla*), Gethi (*Dioscorea bulbifera*), Kulu (*Dioscorea daemona*), Pathal Konhra (*Pueraria tuberosa*).
9. November-December	Fruits	Bar (*Ziziphus jujuba*), Tamras (*Psidium guajava*), Sarifa (*Anona sqamosa*).
	Leaves	Kudrum (*Hibiscus cannabinus*).
	Vegetables	Laoka (*Lagenaria vulgaris*), Radish (*Raphanus cauditus*), Semi (bean).

Month in which obtained	*Item*	*Name of article*
	Roots	Sweet Potatoes (*Amorphophallus companulatus*), Gethi (*Dioscorea bulbifera*), Khania (*Dioscorea pentaphylla*), Aru (yams), Kulu (*Dioscorea daemona*).
10. December-January	Fruits	Chihor (fruit of *Madhuca indica*).
	Vegetables	Semi (bean), Laoka (*Lagenaria vulgaris*), Tomatoes (*Lycopersicum esculentum*), Radish (*Raphanus cauditus*), carrots.
	Roots	Sweet Potatoes (*Ipomea batatas*), Khania (*Dioscorea pentathylla*), Gethi (*Dioscorea bulbifera*), Dura (*Dioscorea belophylla*).
11. January-February	Flowers	Gulgul (*Cochlospermum gossypium*), Jirhul (*Lagerstroemia regina macro carpa*), Kachnar (*Bauhinia variegata*).
	Fruits	Bel (*Aegle marmelos*), Khaith (*Feronia elephantum*), Imli-jo-do (*Tamarindus indica*).
	Vegetables	Semi (bean), Tomatoes (*Lycopersicum esculentum*), Brinjals (*Solanum melongena*), Carrots, Radish (*Raphanus cauditus*).
	Roots	Dura (*Dioscorea belophylla*), Gethi (*Dioscorea bulbifera*).
12. February-March	Flowers	Jirhul (*Lagerstroemia regina macro carpa*), Kachnar (*Bauhinia variegata*), Mahua (*Madhuca indica*).
	Fruits	Bar (*Zoziphus jujuba*), Pipal (*Ficus religiosa*).
	Vegetables	Semi (bean), Brinjals (*Solanum melongena*), Tomatoes (*Lycopersicum esculentum*).
	Roots	Gethi (*Dioscorea bulbifera*), Dura (*Dioscorea belophylla*), Kandas (tubers).

References

Bodding, Rev. P.O., 1940: *Memories of the Royal Asiatic Society of Bengal*.

Elwin, Verrier, 1955: *The Religion of an Indian Tribe*, Oxford University Press.

Gadgil, M. & Chandran, Subash, 1989: *Environmental impact of forest-based industries on the evergreen forests of Uttara Kannada district: A case study*, Final report, submitted to Department of Ecology and Environment, Government of Karnataka.

6

People's Perception and Indigenous Knowledge of Land Resource Conservation in Rajasthan

Laj Pal Bharara

Traditional technologies for conservation

MAN'S struggle for land resource conservation started in antiquity. Until recently most of the success in obtaining higher productivity from arid and semi-arid regions came through the traditional techniques of land resource conservation. The techniques were the outcome of trial and error method and values pertained to drought adjustment mechanisms and land utilisation pattern involving rain-fed subsistence agriculture. Water harvesting for human, livestock and agriculture use like pond catchment and khadeen system (a traditional method of water harvesting) were in vogue. A number of conservationist techniques that mitigate land resource degradation were practised. Thus traditional wisdom in the form of land and water use systems, cropping patterns, livelihood and life-styles, water harvesting, livestock migration and pastoralism, etc. in the form of techniques have evolved to cope with the droughts and desertification and to maximise conservation of land and water to maintain plant, animal and human life on a sustainable basis in the fragile eco-system.

But recently due to socio-economic, cultural and technological transformation of the society and intense human and livestock pressure, there has been a breakdown in the traditional wisdom affecting the balanced use of the ecosystem. This has resulted in ecological deterioration and accentuation of desertification affecting the productivity of the land resources in the region.

Indigenous knowledge

The concept of indigenous knowledge used here is local knowledge that is unique to a given culture or society. It is the basis for agricultural development through land resource conservation encompassing indigenous capacity building practices of regaining fertility of land. Such a knowledge is passed down from generation to generation, usually by word of mouth or by practice. The knowledge has value not only for the culture in which it evolves, but also for scientists and planners from outside. Development professionals increasingly are seeing the value of this type of knowledge in solving agricultural and environmental problems. Because indigenous knowledge systems are adapted to local cultural and environmental conditions, they often are superior to modern technology brought in from outside. Today, because of its oral tradition as well as the introduction of new technologies, preservation of indigenous knowledge is at risk. It is often unknown to or ignored by development workers seeking solutions for agricultural and rural development problems. Solutions offered by a project may fail because they do not fit in with the local knowledge. Indigenous knowledge may suggest alternatives. Methods used effectively by one society could be used to solve problems faced by another. A blend of indigenous and modern approaches may be most appropriate. Many successful innovations are based on indigenous knowledge.

The success of a development project often depends on local participation. Familiarity with indigenous knowledge can help the change agents to understand and communicate with local people. This enables project staff and local people to work as partners in planning and implementing the project. Indigenous knowledge represents successful ways in which people have dealt with their environment. But because much indigenous knowledge has never been documented, it is being forgotten as it is replaced by modern education and technology. As the scientists are trying to preserve the world's rich plant genetic diversity in 'gene banks', it is important to do the same for the world's indigenous knowledge base before it is lost forever.

The focus of this paper is on the implications of traditional system of land resource use for rural development and environmental conservation in the arid zone of Rajasthan. For empirical field investigations, a 'Micro-Level Approach', focusing on each fragment of the cultivator's land resource, was adopted. People's perceptions and traditional wisdom of land resource pertain to its nomenclature in relation to physical characteristics and quality; utilisation, conservation and management techniques of soil and water conservation; traditional knowledge of the environment; and actions and activities relating to indigenous capacity building practices followed.

Findings reveal that the farmers based on generations, old experience have evolved local nomenclature for different types of lands and their characteristics. Traditional nomenclature is associated with the topography, quality and fertility of land; type of crops grown and vegetation available; and water harvesting structures and aspects of social structure. Based on this, the farmers have identified and classified four types of cultivable lands in terms of fertility. Excellent type of lands locally named as Unav, Unam or Kundala are known as Barani I (B-I) and come under land depressions. Very good fertile lands (B-II) or Jod, Jodio are flat alluvial plains. Good category (B-III) are termed as Dabla or Banwla related to vegetation available. Fair category (B-IV) is Dada, Tiba and Dhal characterising sandy, hummocky, duny or slopy nature of land. Poor fertile land is called as Magro or Magrio indicating gravel or rocky nature of uncultivable village common lands including fallow, waste, barren, and Goacher[1] lands.

---

1. Village grazing lands.

Traditional knowledge of the environment

Environmental management also focuses on the traditional perception of the environment by the rural people. This includes not only how the rural people see the environment, but more particularly how they understand it, what they know about the environment and how to exploit it, and how accurate this knowledge is. These perceptions and knowledge become the base of modern technology for environmental management.

PERCEPTION OF AGRICULTURE

Based on experiences, the farmers see the traditional system of agriculture as generally more balanced, producing slower growing crops and lower yields but preserving long-term soil fertility, whereas they perceive the modern system of agriculture as less balanced, producing faster growing and higher yielding crops but at the cost of long-term soil depletion. For example, the traditional system classifies crops, soils, fertilisers and even water according to the humoral concepts of hot versus cold and dry versus wet. This basic principle of the system is to moderate and hence preserve the agricultural environment, by balancing inputs with different humoral qualities. This would mean prescribing hotter fertilisers, (viz., chemical fertilisers) for lighter and cooler soils and prescribing cooler fertilisers (viz., manure) for heavier and hotter soils. This insight into farmer's views yields a number of concrete recommendations for modifying modern inputs to increase the farmer's acceptance of them, e.g., by selecting cooler and wetter crop variation, chemical fertiliser and water.

KNOWLEDGE OF PHYSICAL CHARACTERISTICS OF SOIL

People's perception based on generations, experience revealed five types of rain-fed cultivable lands having excellent, very good, good and fair level of productivity. In addition to this village common lands like Goacher, government fallow lands like Ussar[2],

2. Barren or unproductive land.

Hilly[3] and others were perceived as poor and less productive in terms of productivity and constituted about 13 per cent of total lands. Of the total cultivable lands, excellent quality (Barani I) constituted about 7 per cent; very good (Barani II) 3.5 per cent, good (Barani III) 5.5 per cent and fair (Barani IV) 71 per cent. Colour of soil was perceived as black in B-I category, less black in B-II, yellow in B-III and white in B-IV. Texture composition perceived in terms of proportion of type of soil revealed that B-I land had black and yellow soil (50 per cent) each, while B-II had 40 per cent and 60 per cent respectively, B-III land had yellow soil (20 per cent) and sandy soil (80 per cent) while B-IV land had proportion of yellow, sandy and gravel soils (10 per cent, 40 per cent and 50 per cent) respectively. Thus lands fair in productivity have predominance of gravel and sandy soil, good lands have predominance of sandy soil, very good lands have predominance of yellow soils while excellent lands have equal proportion of black and yellow soils indicating high productivity. People reported that with the decrease in land productivity, the proportion of black and yellow soil goes on decreasing while sand and gravel proportion goes on increasing.

Depth of fertile soil is related with the quality of land. It increases with the increases in quality of land and vice versa. It was perceived to be 135 cm, 105 cm, 75 cm and 45 cm in depth of excellent, very good, good and fair quality of lands respectively. Quality of land is also associated with the topography of land. Good to excellent quality of land was characterised to be of less slopy (20-30 per cent) in nature with more flat area (70-80 per cent) as compared to a fair quality of land having more slope (50 per cent and above) and less flat area (50 per cent).

Moisture holding capacity of the soil is also associated with the quality of land. It increases with the increase in quality of land and vice versa. Retention of moisture in a field in terms of maximum number of days was reported to be highest, viz., 10 days in excellent quality of land followed by eight days in very good

---

3. Gravel or rocky type of land.

Table 1: People's perception of physical characteristics of land resource potential and its indigenous knowledge among desert dwellers

Physical features	*Perception of quality of land resource potential*				
	Excellent	*Very good*	*Good*	*Fair*	*Poor/unproductive*
1. Type of Land	Barani I	Barani II	Barani III	Barani IV	Common village lands
2. Area in *bighas*[4] (% of total)	1000 (7%)	500 (4%)	700 (5%)	10000 (71%)	1800 (13%)
3. Average size of fragment (bighas)	15	60	40	70	Goacher786 Hilly 183 Ussar 13 Other 828
4. Colour of soil	Black	Less black	Yellow	White	White and red
5. Texture composition:					
Black soil	50%	40%	-	-	Black/
Yellow soil	50%	60%	20%	10%	Yellow
Sandy soil	-	-	80%	40%	Sandy
Gravel	-	-	-	50%	Gravel Rocky
6. Depth of soil	135 cm	105 cm	75 cm	45 cm	-

4 *Bigha* is the local unit of land measurement: 1 hectare of land = 6.25 *bighas*, one acre of land = 2.5 *bighas*

Physical features	*Perception of quality of land resource potential*				
7. Slope: - Slopy area (*bighas*) - Range of slope (% & size in cm) - Flat area (in *bighas* & %)	200 20% 60-120 800 80%	150 30% 90-150 250 70%	200 28% 90-150 500 72%	5000 50% 30-90 4000 50%	- - - - -
8. Distance from village core (km)	1-1.5	1-2	2-3	2-4	-
9. Moisture holding capacity (maxim. no. of days)	10	8	6	4	-

type, six days in a good type and four days in a fair type of land.

KNOWLEDGE OF QUALITY OF LAND-CROP RELATIONSHIP

Knowledge of crop-land relationship revealed that the quality of land is positively associated with the type of crops grown and its productivity. In excellent, very good and good quality of lands, Til (*Sesamum indicum*) and Bajra (*Pennisetum glaucum*)are the chief crops grown in rainy season, while in fair quality of lands Moth (*Phaseolus acutifolius*) and Guar (*Cyamopsis tetragonolobus*) are the chief crops. Sewaj[5] and well irrigated lands are available only on excellent type of lands where wheat and Raira (*Brassica juncea*) are grown. Production of crops is highest in excellent, very good and good quality of lands as compared to a fair quality of land in the areas. Based on crop-land relationship, soil is also classified as *meethi* (sweet), *khari* (saline) and *kharch* (absolutely saline). Production of Bajra has shown a decreasing trend with the decrease in quality of land, viz., 12q/ha (q = quintal = 100 kg) in excellent type of land and 8q/ha in fair quality of land.

Production of Mung (*Vigna radiata*) was found to be 10q/ha in very good and excellent type of land and 3q/ha under fair quality or land. Moth showed an increasing trend under fair quality of lands. Whereas production of Til was highest (15q/ha) in excellent and very good quality of lands it was 7q/ha under good quality of land (Table 2).

Revenue charged per unit of land is positively associated with the quality of land in the area. It was 56, 40, 28 and 19 paise per *bigha* of excellent, very good, good and fair quality of land respectively. This shows that the revenue is fixed, according to the quality of land in the area and it increases with the increase in quality of land in terms of productivity and vice versa.

Use of FYM (Farm Yard Manure - dung of livestock) varies with the quality of land and type of crops grown. There are

---

5. Area where rain water accumulates and *rabi* (dry season) crop is grown when the water recedes.

Table 2: Socio-economic characteristics and perceptions of land resource potential

Socio-economic characteristics	*Perception of Land Resource Potential in Terms of Fertility*				
	Excellent	*Very good*	*Good*	*Fair*	*Poor/unproductive*
1. Traditional Nomenclature	Badio/Unav	Kundala/Job/ Jodio	Dabla/Banwla	Bedala/Dhal/ Tila	Magrio/Hills Ussar/ Goacher/Orona
2. Crops Grown	- Mung, Til, Bajra & Jowar (*Sorghum bicolor*) - Wheat and Raira (under Sewaj or well-irrigated)	Bajra, Mung & Til	Mung, Bajra & Til	Moth, Guar & Bajra	Grasses, Shrubs, trees
3. Rotation	Til-Bajra-Jowar-Bajra	Mung-Bajra-Till-Mung-Bajra	Til-Mung, Bajra	-	-

Socio-economic characteristics	*Perception of Land Resource Potential in Terms of Fertility*				
	Excellent	*Very good*	*Good*	*Fair*	*Poor/unproductive*
4. Production (in quintals/ha)[6]					
- Bajra	12	12	10	8	-
- Mung	10	10	8	3	-
- Moth	6	6	3	6	-
- Jowar	8	6	6	6	-
- Till	15	15	8	No	-
5. Revenue fixed (paise[7]/*bigha*)	56	40	28	19	-
6. Number of rains sufficient for production	2	2	4	4	-

6. One quintal = 100kg.

7. Indian currency: 100 paisa = 1 rupee.

three types of FYM being used by the dwellers; cow dung (*gobar*) is preferred most and used in clay sandy soil where Bajra and Mung are grown. Camel manure (*sand ki khad*) is used in sandy soil where Bajra is grown. Sheep and goat manure (*mrgani*) is given last preference on Barani lands. It is mostly suitable for clay-sandy soil of well-irrigated lands where vegetables and irrigated crops are grown.

The number of rains sufficient for crop production varies with the quality and moisture-holding capacity of the land. It was reported that even two good rains are sufficient for production of crops on excellent and very good quality of land, whereas on good and fair quality of land the minimum of rains required for adequate crop production is almost double, viz., four to five good rains for each crop.

TIMING OF RAINFALL INFLUENCES/DETERMINES THE ALLOCATION OF LAND AND THE NATURE AND EXTENT OF CROP SOWN

The timing of rainfall in a particular year affects the planting strategies of that year. Differences in the timing of the monsoon also affect the cultivator's sowing decisions and allocation of land to be cultivated and a particular crop to be sown. The cultivators do not follow any singly strategy: they face high environmental variability by shifting their own strategies to meet the agricultural possibilities (Table 3).

Data revealed that if monsoon starts early in the month of Jeth (May-June) the cultivator initially sowed Bajra on an average 60 per cent of his total cultivable land leaving 35 per cent to be sown later on with other crops and 5 per cent to be left fallow for the livestock to graze and to regenerate fertility. When the rains are in Asadh (June-July), 90 per cent of his land is sown (Bajra on 50 per cent, Moth on 20 per cent, Til on 5 per cent and Guar on 15 per cent) and 10 per cent is left fallow. Similarly during Savan (July-Aug.), 75 per cent of the area is sown (Bajra on 40 per cent, Moth on 23 per cent, Til on 2 per cent, Guar on 10 per cent) and 25 per cent left fallow. If monsoon is late in Bhadon (Aug.-Sept.) hardly 50 per cent (half of the area) is

Table 3: Month of rain and type and extent of area sown with crops

Month of rain	Type and extent (%) of crop sown area					
	Bajra	Moth	Til	Guar	Others	% of total area sown
Jeth (May-June)	60	-	-	-	-	60
Asadh (June-July)	50	20	5	15	-	90
Savan (July-August)	40	23	2	10	-	75
Bhadon (August-September)	-	33	-	17	-	50
Asoj (September-October	-	-	-	5	5	10

sown (Moth on 33 per cent and Guar on 10 per cent) and half left fallow. During very late monsoon in Asoj (Sept-Oct) which is very rare, hardly 10 per cent of the area is sown (Guar on 5 per cent and others on 5 per cent) and 90 per cent is left fallow. If the rainfall distribution is good and effective, the fallow lands are also brought under cultivation. However, there is a trend that with a late start of monsoon from Asadh to Bhadon months, the extent of area sown with Bajra and Til goes on decreasing while the area sown with Moth and Guar increases. Emphasis on growing fodder crops like Guar aims at protecting the village livestock and fallow lands go on increasing. These are the strategies followed by the cultivators who do not sow mixed crops. Thus the traditional knowledge of what can grow, how much area is to be sown, shifting emphasis from subsistence-*cum*-commercial attitude to subsistence given the monsoon condition of each year, provides a base for further decisions.

Soil and water conservation strategies

The traditional system of dryland agriculture encompasses a number of conservationist techniques that mitigate land resource degradation. These include indigenous capacity building practices of regaining fertility of land by fallowing; protecting land against erosion by growing particular crops; stabilising crop production and as a security in risk prone agriculture. Based on these, the various strategies pertain to agriculture, viz., traditional land use pattern: subsistence crops grown and practices followed; rain-water harvesting-storing to prevent run off for moisture conservation; livestock migration and pastoralism as ecological adaptation; protecting forest and vegetation through religion; cultural integration and balanced resource use; sedentary nomads owing livestock; and traditional forms of mutual Cupertino contribute to a sustainable development of the ecosystem.

Indigenous knowledge of following conservation strategies revealed that crop residues, manure and leaves of vegetation are being used for mulching the fields. Growing of legumes, (cowpea, Mung-bean, Moth-bean), creepers as watermelon, Kakkri[8] (*Cucumis melo*), Kachri[9] (*Cucumis callosus*), Tumba (*Citrullus colocynthis*) and fodder crop (clusterbean) provide better protection to land against erosion. Strip cropping is prevalent in irrigated areas. It meets food requirements and also leads to minimum erosion. The concept of crop spacing is fully known to the desert dwellers and is imbibed into the cultural framework of the society. Mixed cropping of cereals with legumes is traditionally practised for stabilising crop production and as a security in risk prone agriculture (Table 4).

Various other techniques include growing of subsistence crops, viz., local varieties of long duration, crop rotation, mixed cropping and mixed farming, manuring but not using chemical fertilisers, cross ploughing, ploughing by bullocks or camels, constructing bunds and leaving post-harvest stubble and seeds in the fields to control erosion.

---

8. The seeds of this cucumber are sown together with the *Kharif* (wet season) crops. The ripe fruits are eaten, sold in the market and also dried for vegetables.

9. This small cucumber grows naturally in a crop field during the rainy season. The ripe fruits are used for eating and for vegetables also.

Table 4: Traditional knowledge and indigenous capacity building practices of different types of land resource potential for sustained productivity in an arid village (Sar)

Indigenous capacity building practices	Types of land resource and its grading in terms of productivity			
	UNAV (Excellent)	JOD (Very good)	DABLA (Good)	DADA/TILA (Fair)
I. Fertility Building Practices				
1. Fallowing of cultivable land	Rare	1-2 years after 4 years, continuous cultivation	1-2 years after 4 years, continuous cultivation	1-2 years after 4 years, continuous cultivation
2. Main cops grown and rotation	Til-Bajra-Mung Jowar-Til	Mung-Til-Bajra Jowar	Mung-Bajra-Moth	Moth-Bajra-Jowar
3. Mixed cropping	Bajra & Til	Bajra & Mung	Bajra & Moth	Moth & Bajra
4. Mixed farming (%, income)	Agro-pastoral-cum-silvi (60-30-10)	Agro-pastoral-cum-silvi (60-30-10)	Agro-silvi-cum pastoral (60-20-20)	Pastoral-cum-agro-silvi (60-30-10)
5. Manuring	FYM; & Chemical fertiliser for wheat Jowar	FYM; & Chemical fertiliser for wheat Jowar	FYM's rare use	No use of FYM

Indigenous capacity building practices	Types of land resource and its grading in terms of productivity			
	UNAV (Excellent)	JOD (Very good)	DABLA (Good)	DADA/TILA (Fair)
6. Leaving crop residues while harvesting	Yes	Yes	Rare	No
II. Soil and Moisture Conservation Practices				
1. Khadin system	-	Yes	-	-
2. Levelling	Nil	Nil	Nil	nil
3. Weeding	Twice	Twice	Once	Once
4. Kana bandha[10]	Yes	Yes	Yes	Yes
5. Dhora bandi[11]	Yes	Yes	Yes	Yes
6. Shelterbelts	Nil	Nil	Nil	Nil
7. Residual leftovers	Yes	Yes	Yes	Yes

10. Field bunds made out of mud and bushes to prevent the run off of water.

11. Bunding or plugging the whole of gully or Nala in a field with mud to prevent the run off of rain water from the field. It is also done to store moisture in the field.

Strategies of a biomass production are: utilisation and conservation including lopping of top feed of trees like Khejri (*Prosopis cineraria*) and shrub Bordi (*Zizyphus nummularia*), cutting of shrubs for fodder; cutting, clipping and collection of grasses; collection of tree (Khejri, Babool) pods for feed; harvesting of crops for fodder; collection and storage of fruits and seeds of Tumba, Matira (*Citrullus lanatus*), Kakkri, Kumat (*Acacia senegal*), Kair (*Capparis decidua*), Kachri, Khejri, Bhrut (*Cenchrus biflorus*), etc.

Culture, tradition, folklore and religion have also played a major role in the development and environmental preservation of the delicate eco-structure of this land. Religious sanctity of Orans,[12] Gouchars, tanks and hills and various other observances and taboos to protect the environment, all formed a part of the complex web of social co-existence and co-habitation with nature.

Conclusion

Traditional strategies have come under strain, due to increased pressure of people on land; side effects of public interventions; institutional and technological changes; increased expectations and a decreased consideration for the interests of others leading to an erosion in cooperative lifestyles and rise in conflicts and other reasons.

In the last few decades of reckless development and overexploitation of resources due to increase in human and livestock population, degradation of resources has taken place. Now it requires to reassess our development approach and to once again restore and improve the time — honoured traditional knowledge systems of environment, common land, water harvesting, catchment conservation and flora and fauna protection. It is essential that these systems are not lost and destroyed but are reinforced both in the social and the scientific language resulting in strengthening the economy of the area

---

12. Village common grazing lands left in the name of a temple where religious prohibition of cutting or lopping of green trees is prevalent.

through resource conservation and sustainable livelihoods. The modern knowledge, no doubt, is needed for advancement. But what is required is a synthesis of traditional wisdom with the modern knowledge to evolve indigenous technologies that would, in other words, be endogenous by the total involvement of the stakeholders.

References

Bharara, L.P., 1960: Social Aspects of Drought Perception in Arid Zone of Rajasthan, *Annals of Arid Zone*, 19 (1 & 2), pp. 154-67.

———, 1978: Social Indicators of Drought Prediction in a Village Community of Rajasthan Desert. Presented at the International Symposium – Post Plenary Session – Anthropology and Desertification, Central Arid Zone Research Institute (CAZRI), Jodhpur.

———, 1986: Indigenous Knowledge. A Coping Mechanism in Drought Prone Areas and Changes and Consequences Accelerating Risks and Vulnerability to Desertification in Western Rajasthan. Paper presented at the National Seminar on Control of Drought. Desertification and Famines, organised by INTACH, New Delhi, IIM Bangalore, and Research Foundation for Science Technology and Natural Resource Policy, Dehradun; held at India International Centre, New Delhi on 17-18 May.

———, 1992: Perception of Drought and Desertification by Indian Desert Dwellers, In: Noor Mohama (ed.), *New Dimensions in Agricultural Geography*, Vol. Six, pp. 67-92, Concept Publishing Co., New Delhi.

———, 1993: Indian Pastoral Communities Following Livestock Migration as a Drought Adjustment Strategy, *J. Anthrop. Surv. India*, Vol. 42, pp. 30-54.

Bharara, L.P. and Mathur, Y.N., 1994: People's Perception and Indigenous Knowledge of Land Resource Conservation, Utilization and Management Strategies in Arid Zone of Rajasthan, India, in 8th ISCO, Soil and Water Conservation, Abstracts, pp. 462-5, Oxford & IBH Publishing Co. Pvt. Ltd., New Delhi.

Bharara, L.P. and Seeland, K., 1994: Indigenous Knowledge and Drought in the Arid Zone of Rajasthan, *Internationales Asienforum*, Vol. 25, No. 1-2, pp. 53-71.

Mann, H.S., Malhotra, S.P. and Ankarnarayan, K.A., 1977: Land and Resource Utilisation in the Arid Zone, In: *ICAR, Desertification and its Control*, New Delhi, pp. 89-101.

Singh, L.P., Singh, A.K. and Singh G., 1994: Farmer's Awareness, Knowledge and Participation in Forestry Programmes. A Case Study, In: Singh, Panjab, Pathak, P.S. and Roy, M.N. (eds.): *Agro-Forestry Systems for Degraded Lands*, Vol. 2, pp. 308-12.

7

Conflicts, Resolution and Institutions in Forest Resources Management

Experiences from the Traditional Mountain Communities of Arunachal Pradesh[1]

Ruchi Pant

Introduction

THE incidences of conflicts over forest resource use have increased over the years in Arunachal Pradesh. Literature review shows that conflicts over use of forest resources in the past in the State of Arunachal Pradesh were insignificant. The abundance of the resource on the one hand and the negligible impact of market forces on the other contributed largely to the absence of conflicting and competing interests over the resources in this state. It is scarcity that induces conflicts. Above all, the severity of the conflict prevention mechanisms in the society has served

---

1. This paper is a result of a short-term research undertaken by the author in 1995 to assess the legal plurality prevailing in the State of Arunachal Pradesh in India with regard to forest management and the efficacy of the institutions present in the state regulating use and extraction of forest resources. The study also examined several instances of disputes and violations of the forest laws, both formal and customary; and the processes of dispute resolution in the State. The original version of the report can be obtained from the Centre for Environmental Law, World Wide Fund for Nature, 172-B, Lodhi Estate, New Delhi 110003, India.

as deterrent to the members from violating the social and customary norms. An increase in the number and a change in the nature of disputes over natural resources utilisation has been witnessed over the last two decades in this State. Commercialisation of resources, especially timber, has also led to an increase in the number of conflicts in the State. The village institutions have hitherto been successful in resolving conflict that arise with regard to forest management. The forest management in north-east India differs from that of the rest of India due to the fact that the forests of north-east India including Arunachal Pradesh were exempted by the British from being nationalised in the late nineteenth century.[2] Hence the ownership and management of large stretches of forest tracts in this region continued to vest in the local communities. These forests have been termed as Unclassed State Forests (USFs)[3].

Arunachal Pradesh

One of the seven States of north-east India, Arunachal Pradesh falls in the Eastern Himalayan Region, a region known as a global hotspot for its rich biological diversity.[4] The state owes its unparalleled biological diversity to its location at the tri-junction

---

2. India was under British rule between 1857 and 1947. The British created the Forest Department in 1864 and drafted the first forest legislation in 1865. The Act saw several amendments in 1878, 1880, 1901, 1919 and finally in 1927 the Act took the shape in which it exists today. This latter Act nationalised all forests in India with some exceptions.

3. Though USF is a legal category under the prevailing Forest Act in the State, yet these forests under community ownership have not been formally notified as USFs. All forests in the state that do not form part of either a protected area such as a national park and a sanctuary or a reserved and village/Anchal forest reserve, are collectively termed as USFs for official convenience. Only timber permits are issued by the State Forest Department in these forests. Harvest and extraction of other forest resources in USFs is governed by customary norms.

4. Myers, N., 1988, Threatened biotas: 'hot spots' in tropical forests. *The Environmentalist*, 8 (3):187-208.

of the Paleoarctic, Indo-Chinese and Indo-Malayan biogeographic realms. Arunachal shares its borders with Bhutan in the west, China (Tibet) in the north and Myanmar in the south-east. It is the largest state of north-east India encompassing a geographical area of 83,743 sq km. The State of Arunachal is endowed with a vast forest cover. While India's forest cover at the national level averages at 10 per cent, for Arunachal the figure stands at 61 per cent. Out of the 83,743 sq km geographical area of Arunachal Pradesh, 61.54 per cent (51,540 sq km) falls under forests of which Unclassed State Forests constitute 62.17 per cent (32,039 sq km).

This rich diversity in the State supports ethnic communities which can be divided into 25 major and over 110 sub-tribes, each having a distinct socio-cultural identity.[5] The total population of the state as per the 1991 census is 8,64,558. The indigenous population, which constitutes 63 per cent, lives in close association with nature and utilises a wide variety of forest resources for sustenance and livelihood.

Arunachal Pradesh has been endowed with diverse and magnificent wildlife. There are over 5000 plants, 85 rare mammals, over 500 birds and a large number of butterflies, insects and reptiles. Twenty-five species of mammals found here are listed as endangered species in Schedule I of the Wildlife Protection Act, 1972 of India.[6]

The rich forests of Arunachal Pradesh comprise the following types: tropical, sub-tropical, pine, temperate and alpine. These varied forest types harbour a wide range of floral species: about 52 species of Rhododendrons, 18 species of Hedychium, 16 species of oaks, three species of conifers and a large number of ferns and lichens. Arunachal can also be known as the 'land of

---

5. In accordance with the requirement of the Indian Constitution 1950, the Presidential Order of 1956 declares all tribes of the North East Frontier Agency (NEFA), now known as Arunachal Pradesh, to be treated as scheduled tribes.

6. *Arunachal Forests 1993*, Department of Environment and Forests, Government of Arunachal Pradesh, Itanagar.

orchids' as it houses over 500 species of the 1000 species found in the country.

ADMINISTRATIVE HISTORY OF ARUNACHAL PRADESH

In the last century Arunachal Pradesh has been under different administrative systems and has had different nomenclatures. During British rule, in 1875, this area was declared as a part of the inner line as provided by the Bengal Eastern Frontier Regulations, 1873, for peace and good governance of the area. As per the Regulations, no British subjects or foreign residents (even Indians) were allowed to cross these lines without a permit from the Executive Officer. In 1914, the area was administered as a part of the North East Frontier Tract and was placed under the charge of Political Officers. The same area, later in 1921, under the provisions of the Government of India Act of 1919, was declared as 'backward tracts' and in 1937 under the provisions of the Government of India Act, 1935, came to be known as 'excluded areas'. In 1954, the area was renamed North Eastern Frontier Agency. With the enforcement of the Constitution of India in 1950, this area came under the realm of Part B of the Sixth Schedule thereof. In 1972, with the enactment of the North Eastern Areas Reorganisation Act, the area was declared as a Union Territory and was accorded the name Arunachal Pradesh. Thereafter it ceased to be a part of the Sixth Schedule of the Constitution. Arunachal Pradesh attained statehood in 1987. Over the years the State has been divided into a number of districts for better administration and these districts have been placed under the charge of Deputy Commissioners.

LEGAL PLURALISM

Arunachal Pradesh is a legally pluralistic state, which implies that several legal systems such as statutory laws and customary laws (social and religious) coexist within a political system.[7] Arunachal is also a unique state, as there coexist two parallel

7. Cathie, J. Willie, 1996. "Comparative Jurisprudence" in Hunter and Whitter (eds.), *Encyclopedia of Anthropology*, pp 84-5.

bodies or institutions dispensing justice in the State. One is the executive or the administration wing, which administers justice in the State through its tiers at different levels. This could be termed as the institution dispensing justice in the formal sector, equivalent to the District Courts, the High Court and the State Forest Department.[8] The other institution is the traditional tribal or village councils. These village councils still exist in most parts of the state and are effective in adjudicating matters within their jurisdiction. Several instances of successfully resolved forest disputes by these councils have come to light and have been documented by the author. The village council could be of different forms. The forms prevailing in the Lower Subansiri District among the Nishis and Apatanis are discussed later. But besides these, different tribes in Arunachal have different forms of traditional institutions. In some tribes such as the Khamptis, Wanchos and Noctes, it is the village or clan chief and his council; among some tribes such as Monpas, it is the elected leader, and yet among others such as the Adis and Apatanis it is a council. The State officially recognises the existence of these traditional forms of institutions and takes cognisance of the decisions of these bodies. The functions of these traditional institutions are not limited merely to adjudication but extend to performing other administrative, religious and social functions as well. These bodies were earlier authorised to try both civil and criminal matters but during the British rule, the heinous crimes were brought to the formal courts, though heinous matters such as robbery are still decided by the former. The traditional institutions rely on customary laws and social norms for conflict resolution. To try heinous offences during the British period, Political Officers and assistants were appointed and special rules[9] were promulgated

8. Forest Officials in the State Forest Department are also empowered to book cases and decide matters related to the violation of the provisions of the Forest Law.

9. The Assam Frontier (Administration of Justice) Regulation of 1945 was extended to Arunachal Pradesh. Section 6 of the Scheduled District Act, 1874 provided for the extension of such a regulation to Scheduled Areas.

for the administration of justice in these areas, as the Civil and Criminal Laws applicable to the rest of the country were prohibited from application to Arunachal. While the Political Officer was granted the powers of District and Session Court, the Assistant Political Officer's powers were equivalent to that of a First Class Magistrate. After Independence in 1947, the posts of Political and Assistant Political Officer were substituted by those of the Deputy Commissioner (DC) and an Additional Deputy Commissioner (ADC), both hailing from the Indian Administrative Service. When disputes between two tribal groups come to the DC or the ADC, often the help of Political Interpreters and Village Council members is sought, as the DC and ADC usually happen to be non-tribals or hail from a different tribe who may not know the customs and social practices of the disputants.

This shows that customary rules have been given due respect in the state and even the formal judicial body of the State resorts to customary law and social practices while adjudicating. It is also evident that Arunachal Pradesh is a semi-autonomous social field — a field that can generate rules and customs having social sanction internally (at the micro-level) and at the same time is subjected to the rules, decisions and forces emanating from an outside source (the national and State legislated laws).[10] It is also learnt from the section on administration that since the enforcement of the constitution, centrally legislated national laws and State enacted statutory laws are also applicable to the State of Arunachal Pradesh. Though the traditional village institutions mainly rely on customary rules while adjudicating, even the DC and the ADC prefer to resolve conflicts keeping in mind the social system and the local customs, except when the matter may be of a serious nature and gravely violates the national law or may have an implication at a larger level.[11]

10. Moore, S.F. (1973), Law and Social Change: the Semi-Autonomous Social Field as an Appropriate Subject of Study, *Law & Society Review*, vol. 7 (4): pp. 719-47.

11. Although no such matters in the field of forest resource management have come to the notice of the author, this opinion is →

Although the British had divided offence into two categories: the heinous and non-heinous, of which the former were to be tried by the Political Officers and Assistant Political Officers and the latter could be tried by the village councils, it has been documented that often even heinous offences were amicably settled by the village while councils the British authorities turned a deaf ear towards these unless these threatened to disturb the law and order situation. The British-appointed POs interfered in cases relating to land grabbing and forcible occupation of fishing-water by powerful villagers which often resulted in long-drawn feuds between tribes or between tribes and villages. Hence often cases relating to murder were dealt with by the village councils.[12]

THE APATANI PLATEAU IN THE LOWER SUBANSIRI DISTRICT

The field selected for this study is the area inhabited by the Apatani tribe, the Apatani Plateau in the Lower Subansiri District. The district lies approximately between 92°40′ to 94°21′ Eastern longitude and 26°55′ to 28°21′ Northern latitude. The district is surrounded by Tibet (China) and some parts of Upper Subansiri District in the north, West Siang district and parts of Upper Subansiri in the east, Papum Pare and the State of Assam in the south and East Kameng district in the west. The Lower Subansiri district covers an approximate area of 10,135 sq km. It has mainly a mountainous terrain, where the hill ranges vary from 1000 to 2500 m above sea level. The heartland of the district is a picturesque plateau at an altitude of 1564 m above sea level, where the district headquarters are located. It can be seen from the statistics provided in the earlier section that the population density of Arunachal Pradesh is low. Hence to study the conflicts and their resolution processes in the state, I felt it best to study

---

→ based on detailed personal communications with the officers in the State during the period 1993-1995.

12. 'A Study of Administration of Justice Among the Tribes and Races of North-Eastern Region' directed by J.N. Das, Law Research Institute, Guwahati High Court, Guwahati, 1986, pp. 312-26.

natural resource management practices among the tribe whose population density is high. The conflicts are more clearly revealed in the condition of resource scarcity. Although the population density for the Lower Subansiri District is as low as that of the State, the population density of the Apatani plateau is very high. The plateau extends over an area of 32 sq km and the people dependent on the resources thereof are 16,335 in number (1981 census). The plateau has been named after the earlier settlers of this area, the Apatanis, although now one comes across many other people living here.[13] The ownership rights and right to resource use (usufruct rights) are limited to the Apatanis though. The other tribes residing in the vicinity of the Apatanis and the neighbouring Apatani plateau within the Lower Subansiri District are the Nishis and Hill Miris.

APATANI

The Apatanis are considered to be the most industrious and enterprising community in Arunachal Pradesh. This tribe is well-known for its wet-rice permanent cultivation-*cum*-pisciculture practices. The Apatanis also use the most effective water-harvesting methods for irrigation. As land is limited, they utilise every inch of land in the most productive manner. The Apatani settlements are normally located very strategically at the base of the forests at the beginning of the paddy fields. All organic waste from the houses flows into the paddy fields along with the rainwater enhancing the fertility of the fields. Since cultivable land is scarce, the Apatanis do not waste land by building huge houses, nor are the houses scattered all over the village area. They are built adjacent to each other neatly in narrow rows. The reason to live in compact villages with houses close to each

---

13. Ziro, the District Headquarter for Lower Subansiri district, is located on the Apatani plateau and was earlier known by the name Hapoli. This place has attracted many outsiders in search of jobs. Since the place is a district headquarter, nearly all government departments are located here. Nepalese and Bihari labour communities find work here in timber trade.

other in the past was also to protect themselves against their aggressive neighbours, the Nishis. Relations have changed over the decades and the belligerent nature of the neighbouring tribes has subsided. In the recent past, as the population is growing, the Apatanis have spread themselves out from the earlier seven villages to 12 villages now. The new village clusters have come up closer to the main road. The Apatani community is sub-divided into clans. Two or three different clans inhabit each village. The head of each clan is called the *buliang*. In the past these *buliangs* represented the highest position in the socio-political, administrative and judicial authority in the state.

NISHIS

The Nishis are also called Nishings or Nishangs. They, like the Apatanis, belong to the Tibeto-Mongoloid race stock. The Nishis are further divided into exogamous clans. Their society is patriarchal and polygamy was prevalent till the last generation. There is a practice of paying the bride-price, ransoms and fines in terms of Mithuns (*Bos frontalis*), a semi-domesticated bovine. The animal is not kept in the homesteads but is left loose in the community forests to graze. The Nishis live in scattered villages and build long houses with several hearths. The number of hearths and as a result the length of the house depends on the number of wives in the family. The Nishis have traditionally been involved in slash and burn cultivation, also known as the *jhum*. This practice is losing its popularity as the number of hands available to prepare the land has reduced with children moving away for education and the decreasing returns from the land. The government is also keen to wean people away from this practice as with the decrease in the *jhum* cycle, this cultivation practice is no longer sustainable and the government feels it is a major reason for deforestation in USF areas. All *jhum* cultivation has hitherto been and is still being practised in USFs. The state is trying to promote permanent cultivation among the Nishis by introducing them to new land uses such as horticulture, tea plantations, etc. This is giving rise to settled cultivation leading to the creation of permanent villages.

The Nishis are not used to living in permanent settlements as their homestead used to move along with the shifting of the cultivation site. They have not held land on an individual basis, but have known only community ownership. The entire forest is considered to be community forest. This new process is leading to a rise in social tension and conflicts. Many people want to find land close to the road side, especially the main road, which is not possible. Other reasons leading to these tensions in the society are mentioned later. No cadastral surveys have been made so far in the state, hence people do not have any documents to prove their rights over land. Now the State Administration has started issuing land possession certificates so that people can avail of the benefits of the government schemes and financial assistance from various institutions, but these certificates are not issued on the basis of any formal land surveys. The process is arbitrary which is another cause for conflicts among people. Moreover, the village elite and stronger tribes with more influence in the power centres have been able to get ownership certificates issued in their name.

Table 1 depicts that 36,210 sq km of the total 51,540 sq km, i.e., 70 per cent of the forests of the State are under the category of USF. 29 per cent of forest is under complete government control as Reserved Forests, Wildlife Sanctuary and National Parks. Another 0.3 per cent and 0.4 per cent are under the category of Village Forest Reserves (VFRs) and Anchal Forest Reserves (AFRs) respectively. AFRs and VFRs are forest categories introduced in the state in 1975 and 1984 respectively with the enactment of the Arunachal Pradesh Anchal Forest Reserve (Constitution and Maintenance) Act, 1975 and an amendment thereto in 1984, viz. Arunachal Pradesh Anchal and Village Forest Reserve (Constitution and Maintenance) Amendment Act, 1984.

Forests

CLASSIFICATION OF FORESTS

Table 1: Classification of forest area in Arunachal Pradesh

Classification of Forest Area (1989-90)	*(Area in sq. km)*
Reserve forests	9,815.37
Sanctuaries	2,526.75
National parks	2,468.23
Classification of Forest Area (1989-90)	*(Area in sq. km)*
Orchid sanctuary	100.00 5,094.98
Village forest reserves	154.92
Anchal forest reserves	256.08
Unclassed State Forests (USFs):	
(a) Proposed reserve forests	3,681.84
(b) Proposed Anchal forests reserve	820.70
(c) Proposed village forest reserve	20.28
(d) Others	31,688.00
TOTAL USF	36,210.86
Total Forests	51,540.00

Source: *Arunachal Pradesh Forest Statistics 1990*, Forest Department, Itanagar, Arunachal Pradesh.

Creation of AFRs and VFRs in USFs is a kind of partnership arrangement between the government and the people, based on the following perceptions of the State Government. Firstly, there is a deeply felt need for conservation and development of forest resources and development of timber trade in the State; secondly, there is a felt need for development and income generation for local people in this difficult terrain, which is characterised by inaccessibility, remoteness, long isolation and lack of alternative economic opportunities. Thirdly, experiences with the self-management of forests owned by the village chiefs of the Nocte tribe in Tirap District under a partnership arrangement with the government since 1948 and administration of a trust fund by generation of revenue from the management of Namsang and Borduria Village Forest Reserve were successful.[14] In order to initiate similar arrangements for USFs, while recognising and respecting the traditional rights of people and communities, VFRs and AFRs were declared in the State. Under this arrangement the State Government meets the ordinary recurring and non-recurring expenditures and also capital expenses for managing the forests. And the net revenue gained every year after meeting ordinary expenses and deducting 10 per cent of the capital expenses shall be shared by the government and the concerned Anchal Samitis/ Village Panchayats[15] in the ratio of 50:50. By carving out AFRs and VFRs from the USF, it is an attempt of the government to bring in USFs under partial control of the government.

Reserved Forests are completely under the control of the State Forest Department. The only category now left with the people is the category of USFs. It is apparent from Table 1 that all forest land that has not been declared Reserved Forest, sanctuary, national park, VFR or AFR, has been clubbed together as USF. There is provision in the Assam Forest Regulations 1891, which is the prevailing forest legislation for the State of Arunachal Pradesh, for the declaration of USF but this process of official notification of USFs has not taken place in Arunachal Pradesh. Hence, USFs continue to be under the management of the local communities governed by customary laws and rules.

---

14. Patnaik, S.S., Barik, S.K., and Shukla, G.P., 'Arunachal Pradesh Anchal Forest Reserve Act 1975, Village Forest Reserve Act, 1984 and New JFM Resolution: A Conceptual Change from Partnership Management of High Forests to Participatory Forest Management' paper presented at the JFM workshop in Itanagar on 3 and 4 October, 1997.

15. With the 73rd Constitutional Amendment, Anchal Samitis and Village Panchayats have attained a constitutional status. These are elected bodies — the latter at the village level and the former at the Anchal level. An Anchal comprises a cluster of villages.

Control and ownership pattern of USFs

The perception of the State Forest Department regarding the ownership of USFs is that it is at the government's disposal and hence under its control. This category of forests has been *de facto* given the term USFs. The State Forest Department has been issuing timber permits for felling timber species in USFs for quite sometime. The local people have not resisted this so far.

The perception of the local people is completely different. They consider all forests within their village boundaries as their own. In several instances in the recent years, even the creation of Reserved Forests is being questioned. The older people in the communities are in full knowledge about the boundary demarcations. The ownership and land use pattern differs from tribe to tribe. For instance, as mentioned earlier, the Nishis have only community forests whereas among the Apatanis, there are three types of ownership over forests. The Apatanis, ownership pattern over land including forests is individual, clan and village or community. The land use of these forests differs on grounds of the ownership pattern. The following table (Table 2) depicts the land use accorded to these USFs in their district according to the ownership pattern.

Table 2: Resource use pattern in USF among Apatanis

Ownership Type	*Resource Use*
Individual	Kitchen garden Bamboo-pine groves Granarysites
Clan land	Burial grounds Hunting ground Wood for house and granary construction Religious and sacred areas
Community or village	Burial grounds Grazing grounds Community hunting for festivals Sites for community worship during festivals

Source: Compiled from: *A Himalayan Tribe: From Cattle to Cash* by Christoph von Fürer-Haimendorf, 1980.

All cultivated land, i.e., irrigated rice fields, dry fields, gardens and bamboo-pine groves, house and granary sites are individually owned. Clan land consists of the sites for public assembly platforms (*lapang*) inside the village, meadow and forest land for use as pasture and burial ground and for hunting and trapping of wild animals. Burial grounds are usually under clan or community ownership, but in some villages clans do not have burial grounds. In such cases, families bury their dead in their cultivated lands by making a special enclosure. Grazing lands belong to the village and cannot be sold or purchased. Kitchen gardens can be acquired by inheritance and/or purchased. These are not along with the homestead but usually located away at the fringes of the fields. As far as hunting is concerned, traps are laid in clan forests but when it comes to community hunting for festivals, it is done in village forests which are located in the periphery of the Apatani country. Clan land is usually not concentrated in one place but is dispersed over hills. Clan members can indulge in trapping on such lands, but other Apatanis are also allowed to hunt, earlier with bows and arrows only, but more recently with fire-arms too. Since hunting by traps is dangerous for others using the forest, within the clan forest, areas have been assigned to individual families so that they can lay the traps in their allotted forest patch. Similarly, individual families are also assigned such areas within the clan forests for extraction of cane. These individuals can sell their right to hunt by trap and to extract cane to other members of the clan owning the forest. With the monetisation of economy and the commoditisation of the forest resources especially timber and cane, conflicts have started arising. Fishing rights in USFs depend on the type of ownership of the forest where the stream or river is flowing.

Individual families own bamboo-pine groves. Although several varieties of bamboo grow in the Apatani forests, not all of these are suitable for house construction. Blue Pine (*Pinus wallichiana*), which the Apatanis believe their forefathers brought along with them when they migrated to this land, is what they grow in the bamboo-pine groves. *Pinus wallichiana* is planted in clan forests also, but then these trees are considered to be private individuals' property and not clan property. Anyone felling these trees other than the owner is fined one cow. The grove, besides growing pine and bamboo, also contains fruit trees and other trees that are considered sacred such as the *thakum*, a plum tree with white flowers. Alienation of Apatani land by their transferance to non-Apatanis is not allowed, although land exchanges hands within the tribe. As cadastral surveys have not been carried out, change in land ownership is difficult to trace. Moreover, no village maps are available.

Dispute prevention and nature of conflicts

Documents available and reviewed show that disputes over rightful ownership, boundary demarcation or resource utilisation were few in the past. There is an adage among the Apatani people which suggests that it is better to forego a piece of land (which includes forest land, cultivable land and bamboo-pine groves) measuring a span of a hand along the boundaries rather than to expend time, energy and wealth over it in cases of any conflict relating to the boundary.[16] In the past once such a conflict became a public concern, it could end up as an expensive affair. The contestants would waste much of their wealth in trying to prove their supremacy and dominance over the other. Some of these practices are no more in vogue.

The *buliangs*, as clan representatives, in the absence of a centralised authority exercising control in the society, used to settle disputes. But their duties were those of arbitrers rather than that of a judge. They usually did not take action unless a dispute became a public issue or else the complainant sought their help. The nature of dispute in the past was limited to theft. Theft of Mithun was a common complaint. Customs, religious and social sanctions and taboos form an intrinsic part of the society and serve as the unwritten laws of a traditional society.

---

16. Personal communications with village elders of Hang village during my fieldwork in 1995.

The following customary practices and beliefs of the Apatani people have hitherto served to prevent disputes.

Other traditional social practices and controls

DAPO

This term has had different meanings at different times. Dapo began as a peace treaty in times of the earlier settlement of the Apatanis in this country. As they were surrounded by the hostile Nishis, the Apatanis wanted to reduce their inter-village disputes to a minimum. The Apatanis are of the opinion that formal treaties of friendship between the villages, which formed a fundamental part of their political system, assured peace.[17] The same meaning continues to be attached to this instrument of peace, though an element of warning has been added to it. In any field or forest area, if there is a likelihood of a dispute arising, the structure depicting a *dapo* is erected at the boundary of such areas to show that this belongs to a particular group so that there is no disagreement on this account. The structure of *dapo* is made out of about three to four feet long bamboo poles and stuck on land with a trophy of a Mithun head on it in the centre. Usually three poles are placed in a vertical criss-cross manner. Nowadays as there are educated people in the region, they have also started writing a note on a wooden plaque that is attached to the centre saying that a fine or punishment would be imposed if the rules of the treaty were violated. While passing through villages one often gets to see these structures at the village boundaries.

BUNING

A society which is prone to complex inter-tribal disputes manages to integrate itself through the *bunings*. *Buning* is a special friend. Von Fürer-Haimendorf has called them 'ceremonial friends'. This concept of *buning* started ages ago and continues to exist. A *buning* relationship can be entered into and can be inherited from one's father. Only men make bunings and their

---

17. *Supra* note 15, p. 129.

wives are obliged to recognise the relation and accept the duties of a *buning* in the absence of her spouse. *Bunings* are made after a long period of friendship and the relationship is given a formal status by inviting *bunings* over to feasts at the time of the Mloko festival.[18] The relation is considered severed when one's friend does not invite one to this festival.

A *buning* has to represent his friend in his own clan and village. If ever bargains have to be stuck in trade or disputes are to be settled, then the *buning* has to do his best to arrive at the best deal for his *buning* friend from the other village or community. *Bunings* are usually chosen from other clans or villages and sometimes even from a different tribe. The purpose of entering into such kinship ties was to close gaps in the social network.[19]

ORDEALS

Some of the dispute settlement mechanisms in the past, which have continued till recent times, especially in the remote areas, include the system of ordeals. If the *buliangs* and the kinsmen were unable to resolve a dispute by negotiations and mediation, the practice of ordeals was resorted to. The ordeal was organised by a Nishi priest called *Nyibu*. The ordeal was deemed to be an appeal to a superior authority, the supernatural power; and the judgement was accepted. One ordeal entailed boiling of water from one of the major rivers of the Apatani area in a section of a bamboo pole. A small stone was put into it and the accused was asked to take out the stone from the boiling water with his bare hands. If the hand swelled in a little while, he was adjudged guilty, else he was innocent. Another ordeal required a piece of heated iron to be placed on the palm of the accused. Three leaves of the saro-tree were to be placed on the palm before

---

18. *Mloko* is a spring festival performed in each Apatani village once in three years.

19. *Mloko* is a spring festival performed in each Apatani village once in three years, p. 127.

placing the iron piece. If the palm showed signs of burns, the accused was pronounced guilty. These practices were common to find out thieves in the society. Such crude practices kept disputes to a minimum in the society and acted as dispute prevention mechanisms. In addition to these social controls, there were other restrictions imposed by the society on themselves which de facto acted as checks on over-extraction of the forest resources. Many elements and resources of nature and environment were either worshipped or else were used in the worship process. Certain species of trees were considered to be sacred while some fruits, trees and animals were important as they held a special position in their rituals and festivals. Hence, controlled utilisation of such species was important which led to the conservation of these species. For instance, in the Mloko festival which is celebrated by a group of villages together and in a rotation, only a paw or a monkey skull is required. In such a case only one monkey would be killed in a year, and this preserved the monkey population. The availability of the paw or skull of this monkey is essential to the celebration of this festival.

The sacred sentiment of people towards certain species of plants is also due to the site where they are planted. Some such places are regarded as the abode of deities. Also sometimes a species is protected for its great value to the society. For instance the *sal* tree (*Shorea robusta*) holds a cardinal position in the social and economic life of a Santhal tribe. The leaves, fruits, flowers, bark and twigs have different uses for the Santhals of Bihar; hence they preserve the species in 'sarnas', their sacred groves. This area is protected to the extent that not even a fallen leaf is picked up from here. Similarly, Apatanis and Nishis protect some trees. The Apatanis normally consider the ancient pine trees planted by their ancestors and forefathers as sacred. Even decaying trees of this species are not felled for private and commercial use. Instances have come to light where fallen sacred trees have been used for the purposes of community welfare, e.g., planks are used as walks to cover marshy and muddy patches on dirt tracks during rains. Nishis have a tree called saro which they consider, as inhabited by their deity. Nishis attach a taboo to the felling of this tree and if anyone violates this religious sentiment, it is the belief of the local people that the deity punishes the offender by causing a death in the immediate family. Flowers of the plum tree, *thakum*, are used in rituals by Apatanis and hence it is considered as a sacred tree.

The tribes in Arunachal Pradesh also maintain periods of prohibitions during which people are restricted from entering the forest for cutting firewood or extraction of other resources or from going to their fields. This is mainly during the festival time and during special religious ceremonies held at homes. No one from the house is allowed to leave the premises of the house during the prohibition period that extends sometimes from three to seven days.

The two main festivals of Apatanis are the *Morom* and the *Mloko*, both of which are associated with the agriculture season. *Morom* is celebrated in winter and is more of a family celebration whereas *Mloko*, which is celebrated in March, ensures the participation of a group of villages. On this occasion, people and friends from other Apatani villages that are not celebrating the festival that year are invited to share the festivities and to join them in the feasts that form a major part of the festival (as mentioned earlier, *Mloko* is celebrated in rotation each year by a different group of villages). This festival is spread over a period of one month. *Bunings* from other villages and tribes are also invited on this occasion to enjoy the celebrations. They are welcomed with rice beer and good food and given gifts on their departure. These are occasions when *buliangs* from different villages meet and renew contacts. The informal meetings provide the *buliangs* with an opportunity to discuss and iron out social tensions that are likely to take the shape of disputes in the future. The *buliangs* sit together when inter-village or inter-clan disputes are to be settled. During the *Mloko* festival, a period of *anyodo* is observed and it is considered taboo if people from villages celebrating *Mloko* go to the forests during the *anyodo* period. *Morom* engenders the fertility of man and fields and is celebrated at the eve of the new agricultural year by sacrificing Mithun and feasting friends and neighbours.

Such restrictions are observed after the completion of any rites and rituals. Performance of private rites are common in the villages and these take place quite often, as part of ceremonies such as weddings or funerals, or reasons as prosaic as an illness in the family, commencement of a house building, a fire in the village or a personal crisis. The Apatanis perform a seasonal rite in July/August in the name of *Yapun*, the god of thunder. The performance of this rite is believed to ward off the danger of hailstorms damaging the crops. No villager is allowed to go beyond the cultivated areas, i.e., to the forests for 10 days after the performance of these rites. Breach of these rules is believed to lead to hailstorms. These rites and restrictions are followed even today.

These practices, to a great extent, have helped in the conservation of natural resources though in recent times, people have found ways and means to circumvent these restrictions, which are affecting a sustainable harvest. For example, with the advent of hand saw mills,[20] extraction of timber, on days other than those of restrictions, has increased. Also, in recent times, the locals have started employing non-tribals to do the extraction on days of restrictions.[21]

Yet another system which has helped to maintain solidarity and unison among the members of this society is called *nago*. *Nago* is a shrine used by a group of clans as a common ritual centre. Sometimes one *nago* is shared by as many as 600 people and this is what lends these clans unity and keeps incidence of disputes low.[22]

---

20. This too may sound like primitive technology to some, but to the Apatanis it is relatively new and it is quite effective for their purposes.

21. Labour from Nepal and Bihar is available in the region now.

22. *Supra* note 15, p. 82.

Many of the social and religious value systems, of which the natural resource conservation formed a consequence, are eroding in the present days and taking its toll on the resource and the social set-up in the society. Number of disputes has increased and the nature of dispute has also changed over time. The factors related to the new situation are discussed in the next section.

Disputes in recent times

It is felt that the social practices were strictly adhered to in the past, which was a reason for so few conflicts becoming public. Disputes were usually resolved with the aid of negotiations by the kinsmen. If the conflict was between persons of two villages, *buliangs* of the villages intervened. If unanimity was not achieved, *gambus* or *lisudus*, whose role is discussed later, were staged. And more recently with the active involvement of the administration, litigants have started approaching the DC. But this happens normally in cases where the litigant is aware of the fact that the decision of the traditional institution will not be in his favour.

While discussing disputes or violations of the social norms with respect to forest and wildlife in the earlier times, the following is the nature of disputes: primarily trespassing in others' forest; area, such as hunting in another's clan forest; laying traps to capture wild animals in the other's forest; stealing of a trapped animal from another's trap, etc. The frequency of disputes was very much less in earlier times when the adherence to social sanctions was strong. Over the last two decades the number and frequency of disputes have increased. The factors that have contributed to this rise in the number of disputes or violations from the societal norms are discussed below.

COMMODITISATION OF THE FOREST RESOURCES, ESPECIALLY TIMBER

As long as the British Indian Inner Line Permit system mentioned earlier was strongly adhered to and very few outsiders entered the region, the communities in the region had a subsistence economy largely based on barter. After Independence with the advent of Indian Administration, outsiders were seeking employment and trade opportunities in the region, and even the system of Inner Line could not stop them. This gradually increased the influx of external influences into the state, which gradually moved towards a monetised economy. New and luxury products entered the market and the virgin forests attracted the timber merchants from outside the state who knew the value of these forests. These merchants not being from the state had no kind of attachments to these resources. With the introduction of new products such as TV, video, motorcycles luxury cars, and came the demand for these items first from the elite in the society. The easiest way to get cash was to sell off timber permits to timber merchants who had started setting up saw mills and plywood and veneer mills in the state. In partnership with the local people mainly politicians circumvent the regulation that no outsider can set up business or buy property in the state.

The younger generation, although aware of the dependence of the community on the forest resources, tends to ignore the conservation value attached to the forests in lieu of the easy gains they see in selling off timber permits or even entering into illicit felling of timber (which means that if the permit is for two trees, they may fell five). A tree-permit obtained from the Forest Department is worth anything between Rs. 6000 to 60,000 (US$150 to 1500) depending on the species indicated in the permit. This has led to violations of the regulations both customary and statutory. It is also important to note that the State of Arunachal Pradesh is solely dependent on the timber industry for earning revenue. Hardly any industry or value addition opportunities other than saw and veneer mills are available in the state. The educated and uneducated youth is largely unemployed.

INTRODUCTION OF NEW ACTORS

Besides the introduction of the timber merchants, the extension of State control over forests can be regarded as one of the reasons for increase in the number of conflicts. Earlier conflicts were between clan members or between villages, but now there are new actors. Both the State as well as the community and individuals are asserting ownership and management roles. Within this arena fall a number of disputes as the State Forest Department declared Reserved Forests in USF (which were under clan or community ownership), thereby extending government control over these forests. People were unaware of this process earlier and the implications thereof, but since the time realisation dawned upon them that the government had notified their forests as Reserved Forests without following the due legal procedure, disputes alleging illegal encroachment upon their forests by the government have started reaching the formal judicial system. People do not allow the forest officials to enter such forests and there is much tension at times, which has led to incidences of violence in the region. In a particular case, one clan in the Apatani area has challenged the declaration of the Tale Valley Reserve Forest[23] and has alleged claim to part of the Reserved Forest area being their clan forest and that the due procedure as mentioned in the Assam Forest Regulations of 1891 for the declaration of Reserved Forest was not followed.

The other trend observed with regard to encroachment disputes is the government allegation that people have cut down forest plantations and encroached upon government forest land. Hence, the Forest Department has issued eviction notices to the local encroachers and they have appealed to the High Court in Guwahati[24] seeking a stay order on such notices. The Court has been extremely favourable to people in such cases and in two cases has granted them a stay and a show-cause notice to the State Forest Department.

---

23. Three years ago this area had been declared as a Wildlife Sanctuary.

24. As there is no High Court for the State of Arunachal Pradesh, people have to approach the High Court of Assam located in Guwahati with their grievances.

URBANISATION

Many Apatanis gave away their prime lands for the establishment of the district headquarters at Ziro in the Apatani area at a very nominal price and in several cases even without receiving any compensation. Although the process of acquisition of land for this purpose took place over a decade ago, people have now come to realise the value of that land. This is bringing about resentment among the people who have lost their rights over such lands without being suitably compensated. Also, with the alienation of this land, pressure on forest land increased as people encroached for cultivation. Those people, who had given up land for the development of the district headquarters, had to look for new land for cultivation now. As land is limited, people had to clear forests to bring more land under cultivation.

INEQUITABLE BENEFIT SHARING

Tree-permits and their sale to other parties has led to an iniquitous accrual of benefits to some influential individuals and clans in the region. This is leading to tension in the society. The Forest Department issues tree-permits only for certain uses.[25] As per

---

25. The Department has often changed its policy from time to time and from district to district. In 1981, the Government of Arunachal Pradesh introduced the village quota system wherein tree-permits were issued for a village's economic upliftment and development. The village councils were asked to nominate persons to work on the trees allotted to the village on behalf of all villagers. This system was revised and tree-permits were issued to individuals too, on different grounds: one being medical grounds and another being education. Since the people do not have enough monetary wealth and there are very few hospitals with all kinds of facilities, people have to travel to distant places to avail of medical facilities outside the State for which they need money. Hence, on production of a certificate from a doctor, tree permits may be granted to an applicant. This is the beginning of a new kind of racket in the State. The number of tree-permits to be distributed in a forest range for operation in a particular year is determined according to the estimated surplus yield of a particular USF, which may vary

→

the norms laid out by the Department, every district is allotted a quota for the year and the Divisional Forest Officer (DFO) is authorised to disburse the quota in his division. These disbursements are often politically influenced and a select few benefit from such permits. The needy are often ignored in this system. One of the drawbacks of the tree-permit system is that an aspirant of a tree-permit may not get a permit to fell a tree from his own USF but from the USF belonging to another clan. Sometimes, the permit may be for another district. This has led to conflicts between the tree-permit holder and the owner of the USF.

UNDERMINING TRADITIONAL VALUES

Many factors have led to the erosion of the value system prevailing in this society. The sanctions of the society have weakened in their effect. This has been observed both among the Nishis and the Apatanis. The influence of the modern education system and the gradual influx of foreign religions has weakened the belief of the local people in their age-old culture and traditions. Although entry of any religion other than the indigenous beliefs is banned in the State, Hinduism and Christianity have made inroads here. Areas and trees that were considered to be sacred are no longer spared from the saw. The fear of the supernatural is waning. Taboos of felling trees during certain periods are now being overlooked by many and losing their meaning under the influence of the new education system. Hence, one finds this

---

→ from year to year. One often wonders how efficient the system is, because many USFs have not even been surveyed properly and annually. Timber permit operation in USF area is allowed only after advance payment of 100 per cent royalty by the permit-holder on the estimated value of the tree as per local volume table which normally varies from Rs. 15,000 to 20,000 (US $ 400 to 500 approximately) for a tree. The monetary involvement being substantial, the timber permit operation is always allowed in the name of an individual in order to ensure collection of government revenue in time and to guard against default of probable permit holders.

especially among the younger generation. This has led to many problems in the society. *Re*, the practice of community farming, has observed a setback as Christians abstain from work on Sundays. Sundays are now spent as rest days when the newly converted Christians go to the church. Hence they do not go to the fields to participate in *re*. Practices such as *lisudus* and *gambus* have lost their significance nowadays. Christoph von Fürer-Haimendorf had called these practices 'safety valves', through which pent up ill-feelings between groups could be discharged with a minimum harm to the tribal community. He has attributed the increase in the number of murders as a result of land (including forest) disputes to the abolishment of these traditional safety valves.[26] The State administration has tried to do away with the system of ordeals in deciding heinous criminal matters which now go to the DC for hearing.

MIGRATION

Unlike the Apatanis, who have practised permanent cultivation in their small valley, space was more than limited for the neighbouring Nishis. The Nishis have practised shifting cultivation for centuries. The term itself denotes that they did not cultivate always the same site, but with the exhaustion of the nutrients of one area, the community would shift their cultivation to another. They would move their homesteads along with the cultivation following a periodic cycle and would return to the same area which was cultivated by them earlier after a fixed number of years to replenish the depleted nutrients. The family which had cultivated a particular plot of land earlier had a right over that land. This family had privilege over the other families to cultivate the same plot of land on returning later and had rights over the produce which grew in the meantime. This is where conflict would often arise, because several or more than one family would claim rights over the same plot of land.

A more recent trend has been towards promoting permanent settlement among the Nishis. This has led to a new problem as

26. *Supra* note 15, pp. 141-3.

Nishis have now started migrating from the interiors to the roadside. Road linkages in the State are limited. This has led to friction between the earlier settlers and the new migrants from the interior. Initially there was enough land but with more and more people opting to have homesteads close to the roadside, there is not enough land to accommodate them all. There are not enough resources, especially forest resources, to accommodate everyone. But this did not stop people to continue to come down to the roadside, which led to encroachments and complex social problems. When there were enough resources, migrants were welcome. They had to arrange a feast (*nyok nemo*) for the community and sacrifice a Mithun to be accepted by the earlier settlers and to acquire rights in the forest area.[27]

CHANGING PATTERNS OF LAND USE

Forest land is being diverted to non-forestry purposes in the area. A few elite families have raised orchards and plantations of tea and cardamom. Some have resorted to monocropping of oilseed crops such as mustard, which has met with resistance by others in the village. The Reserved Forests under the control of the Forest Department, as mentioned earlier, have been carved out of USFs. Though these reserved forests were created more than two decades ago, people still believe these to be under their control. Many conflicts arise when the Forest Department officials in pursuance of their duties such as surveys and markings want to visit these areas and are not allowed by the locals to do so. The Forest Department has persuaded local leaders to grow pine plantations on Anchal and Village Forest Reserve areas in the Lower Subansiri District. These changes in land use have led to conflicting situations in the district. The Forest Department in its efforts to bring USFs under the Anchal Forest Reserve and Village Forest Reserve category are able to convince and elicit the support of the village leaders or tribal leaders and not of the community. The Forest Department requires the signatures of

---

27. Personal communication with Bengia Tolum, a Nishi, who runs a private school in Yachuli, 1995.

the village leaders on the agreement papers. The problems start surfacing at the time of harvest of the resources and reaping of the benefit. This at times leads to inter-village or even inter-clan conflicts as there may be a dispute regarding the boundary of the said forest. The department does not bother to explain to the entire village and seek their support in the matter. The tribal community is not a homogeneous group and has their own power stratification within the society.

OTHERS

Apatanis have started moving out in search of jobs. Those in government employment may not necessarily be posted in the Apatani region. Long periods of absence from their lands are also posing problems and there are possibilities of conflicts arising among brothers and close families. Many young people go away for higher studies or are even away for schooling in certain cases, hence these people are not aware of their entitlements over forests and other lands. They are not even aware of the boundaries demarcating clan, village and individual forests as the boundaries are not marked by any posts or pillars, but by mountains, streams, ridges and valleys, the knowledge of which is generally handed down by the senior members of the family and clan. Land in the Apatani valley being limited, it cannot support the increasing population. Apatanis being good entrepreneurs with business acumen have moved closer to the State capital, Itanagar. They have started acquiring land here at low prices from the indigenous people of this area, mainly the Nishis. This is a cause of resentment and friction between the Nishis and Apatanis.

Dispute processing in Arunachal Pradesh

As mentioned earlier there is a plurality when it comes to the types of institutions involved in dispute processing in the State. On the one hand I have mentioned that there still exists the traditional system of dispute settlement. The system differs from tribe to tribe, but the concept of justice in a tribal society is the same and is based on restitution. Earlier justice and punishments were based on retribution, which sought to restore the balance by methods that were different from those adopted by the developed society.[28] If the disturbance was caused by homicide, killing any member of the clan to which the murderer belonged could restore the balance. Fines, compensation and restitution have now replaced the concept of retaliation. Restitution was made in the form of compensation paid in kind. Mithun was the most common animal by which compensation was paid in the non-monetised society. The essential point according to custom was to assuage the feelings of the aggrieved party. Hence all disputes, when settled, ended in a feast, wherein both the disputant parties solemnly affirmed to bury the hatchet and banish the urge of retaliation. Earlier offences related to forest resources included theft of Mithuns, disputes over fishing rights, dispute over boundaries of bamboo-pine forest and *jhum* land.

In the tribal societies of Arunachal Pradesh, people prefer to settle their conflicts themselves or with the help of kinsmen. Only when a dispute becomes public or the offence committed has a larger implication, the village institution interferes. Normally, it takes up a matter when approached by a disputant. If the village institutions, may it be the *buliang* of the Apatanis or the *Nyels* of the Nishis, were not able to decide unanimously, ordeals were arranged and the matter was left to be decided by the supernatural powers. The other practices in the earlier days were the staging of the *lisudus*[29]

---

28. *Supra* note 11, pp. 273-9.

29. If the dispute between two men of equal wealth and status has been defied a negotiated solution, the adversaries try to shame each other into submission by a *lisudu* competition that involves the ritual destruction of wealth. A man who challenged a co-villager to a *lisudu* competition started by killing one or several of his Mithun in front of his opponent's house and leaving the meat to be distributed to the relatives of his opponent. Sometimes other precious possessions such as ornaments and bells were also smashed. If the opponent accepted the challenge he had to slaughter at least the same number of Mithuns and destroy property

→

and *gambus*[30]. These practices are not encountered in the Apatani area anymore. Maybe in the remoter areas of Arunachal some of these practices are still followed.

With the active involvement of the Indian Administration, the disputants have now an opportunity for forum shopping, which means that the aggrieved party seeking justice has now a choice to approach any institution it wants. The party can either approach its tribal council which could be in the form of the tribal chief, the arbitrers, the elected council, etc. The party also has the option to approach the District Administration. The matters that come to the DC are first attempted to be resolved by the village councils, failing which the DC decides the case resorting to the customary laws. For this purpose the political interpreters who are well-versed with the customs assist the DC by interpreting the customary laws and social practices of the tribe. Sometimes the DC calls the village council and instead of resolving the matter, he asks the village council in his presence to adjudicate and often matters are resolved in this manner. The decision is then documented and signed by the DC. Many

---

→ of equal value in front of the challenger's house. The man who could continue the destruction of wealth for a longer time won thereby his opponent's remaining property in land and movable possessions.

30. *Gambus* were armed contests in which supporters of both litigants turned out for an open fight. The disputant themselves did not participate as they had to be alive to pay the compensation for the casualties that would take place as a result of the combat. As soon as there were one or two fatal casualties on either side, the *gambu* was called off. Von Fürer-Haimendorf says that in the heat of the fight some damage was done to the bamboo-pine groves and gardens, but there was never any large-scale destruction as used to occur in the raids of the Nishis. The fighting in a *gambu* was organised and confined to long-distance arrow shooting and spear throwing. Ursula Graham Bower in her book *The Hidden Land* shows that *gambus* did not involve actual fighting but consisted only of armed demonstration to exert pressure on the opponents.

disputes relating to forestry have been amicably settled in this way.

Of late, there have been some instances where the aggrieved party has by-passed these institutions in the State and has directly approached the High Court alleging violation of fundamental rights and sought a public Interest litigation in the matter. This is observed in matters relating to challenging declaration of Reserved Forests and seeking a stay in encroachment cases. There is also an opportunity for those who first approach the village councils but are dissatisfied with the decision of the councils or *buliangs*. They can appeal to the DC.

ANCHAL SAMITI AND GRAM PANCHAYAT MEMBER

More institutions have entered the fray. In 1967 Gram Panchayats were introduced to reinforce the system of decentralised local self-government all over India. Although such a system already existed in Arunachal Pradesh in the form of powerful village authorities, the government little realised the mistake it was committing. The members of these bodies were to be elected on the basis of 100 adults electing one Panchayat member and every 500 adults electing one Anchal Samiti member. The latter was more powerful, as an Anchal comprised of a number of villages and an Anchal Samiti member had more authority than a Gram Panchayat member had. Since these bodies are politically affiliated, they often have a stronger voice in deciding matters.

Decisions in the two judicial systems and their acceptance in the society

The following section is an attempt to highlight the efficacy of the traditional institutions in dealing with local problems in relation to forest management. Some cases are referred to illustrate the success of the village authorities in resolving conflicts and how these institutions have also been able to tackle the changing nature of disputes and forestry problems.

CASES

1. In this case, *X* belonging to clan *A* from a village struck a deal to sell a part of his land to a government agency to setup an office. While selling part of his own land, he clandestinely included a part of *Y*'s clan forest land (in the deal). *Y* belonged to another clan *B* from another village but his land was adjacent to *X*'s land. *Y* was not in town for some time. People from clan *B* heard about the deal. Clan *C*'s land was also adjacent to *X*'s land and this gave rise to a fear in the minds of the clan that *X* may play foul with them too. So clan *C* also got involved in the matter. *Buliangs* of all the three clans got together and unanimously decided upon the matter. An agreement was entered into and a dapo (peace treaty) was signed which included a warning. A mithun was sacrificed and some rituals were performed to wind up the case. The Mithun was the price *X* had to pay for the offence committed by him.

2. A conflict arose despite a *dapo* being in effect. A labourer working for a timber-contractor started extracting cane from another village's forest while on work. An agreement existed between the Village Volunteers Force of Chulu village and the contractor from Posa village and under this agreement, the boundary between the two villages was clearly identified and a warning was also issued that in case of a violation, the contractor will have to bear the consequences. The contractor from Posa village encroached upon the forest resources of Chulu village. This infuriated the Village Volunteers Force of Chulu and they beat up the labourer mercilessly. The extracted cane was seized. The contractor could not take any action against the Chulu Volunteer Force as he had violated the *dapo* (treaty agreement). The violation took place on 13th July 1995. The *Nyel* got together on 6th August, but as the labourer was still in hospital and could not depose before the *Nyel*, the second and final

sitting of the *Nyel* took place on 8th September 1995, where the Chulu village decided to forgive and forget the wrong deeds after giving another warning to the contractor. The contractor was allowed to take possession of the seized cane on making full payment of the cane extracted. The *Nyel* comprised of elders from the two villages concerned and the mediator was from a third village called Peni.

Note: This is an example of speedy trial and the *Nyel* decided to let the contractor take possession of the forest resource despite the offence, as they did not want the resource to go waste.

3. A complaint was lodged by a *gaonbura* of Kalung village with the DC on 16th May 1994 that his community forest, namely, the Salaya forest was damaged by two persons of Reru village on 14th May 1994. The Extra Assistant Commissioner (Judicial) on behalf of the DC directed two Political Interpreters and two *gaonburas* to proceed to the Salaya forest to assess the extent of damage and to submit their report. The report was submitted on 26th May 1994. Before even a single hearing could take place, a petition for the withdrawal of the case was filed with the DC informing the court that the case was mutually and traditionally settled outside the court. The court gave the order that the dispute was amicably settled and the case would be treated closed.

4. A complaint was filed against *X* in the DC's court for an attempt to murder *Y* on 4th February 1994 with a *dao* (a locally made weapon like a knife). *Y* alleged that *X* had encroached upon his pine garden in violation of an agreement drawn in presence of clan elders on 15th January 1989. A summon was issued to both the parties to come to the court on 18th April 1994. Clan elders got together and amicably settled the case outside after

laying down certain conditions. A fine of Rs. 1000 was settled to be imposed on whoever violated the set conditions. The case was withdrawn from the court.

5. In a case of over-extraction of cane from a Nishi community forest, the *Nyel* took the decision that cane from the community forest can only be extracted for personal *bona fide* use and not for sale.

6. In another incident in the same Nishi village, a person had started cultivation of mustard in a degraded USF; he envisioned this as a business venture by selling mustard seed oil. The matter was raised and a *Nyel* was called. The *Nyel* decided that mono-cropping is harmful for the health of the forest. And the degraded forest should be set aside till there is natural regeneration, as a mixed forest would be best to maintain the ecological balance of the forest.

   Note: The decision of the indigenous institution in the fifth and sixth case shows the wisdom of these people and their collective judgement. The two cases mentioned above are typical of new kinds of disputes that arise in USF. These examples illustrate how the traditional institutes are effectively handling such problems. Although these people are ignorant about the provisions of the national laws, yet their decisions reflect their wisdom in the area of forest ecology and the concept of sustainable utilisation of forest resources.[31]

---

31. The Indian Wildlife Protection Act, 1972, the provisions of which extend to the State of Arunachal Pradesh too, prohibits picking, uprooting, etc., of specified plants. But it offers an exception to the indigenous people to pick or uproot from the forest in the district in which they reside. But this extraction is only limited to *bona fide* personal use and not for commercial purpose and also to a limited number of species. Hence we correlate this with the decision of the *Nyel*. Similarly the decision in the sixth case shows the understanding of the local people of forest ecology and the →

7. A writ petition was filed in the Guwahati High Court challenging the declaration of the Tale Valley Reserved Forest. The petitioner alleged that part of the said forest is his clan forest and the government did not follow the due legal procedure while notifying the forest as Reserved. In this case,[32] the petitioner, Padi Lailang sought quashing of the notification declaring Tale Valley as a Reserved Forest. The court disposed off the matter directing the Government of Arunachal Pradesh for a fair settlement of the case, yet the dispute remains unresolved till date. The boundary issue could never get resolved and in the years that followed resulted in the filing of another matter involving his grandson Padi Richop.

8. Padi Richop of the Padi clan cleared a forest area of about 10 sq km altering the land-use of it. He started a school, a cardamom and ginger plantation, an orchard, a tea-nursery and attempted to settle some local people and some from outside the state who were working as labourers for him. A conflict arose over the ownership of this forest area. Richop alleged that the forest was his clan forest but, at the same time, a group of Nishis were also claiming the same area to be part of their community forest. In 1992, the Panchayat elections were to take place. This new settlement that Richop developed and promoted comprised of 500 persons, hence the area was now entitled to one Anchal Samiti member. Padi Richop was interested in contesting the election of the

---

→ decision is very much in tune with the provisions and objective of the Forest Conservation Act, 1980, which provides that no non-forest activity can take place in a forest area. And plantation has been considered as a non-forestry activity for the purposes of this Act.

32. Padi Lailang *vs*. Union Territory of Arunachal Pradesh, Civil Rule No. 944 of 1983 and Misc. Case No. 513 of 1983, Guwahati High Court.

Anchal Samiti and he stood unopposed from this constituency that he had developed. Now, Nani Ribya, hailing from another Apatani clan, could not accept this and he was successful in instigating some people from a Nishi community who had also been claiming part of the forest as their community forest. The Nishi community flared up and 5-6 trucks full of Nishi students went and demolished parts of the settlement. Padi Richop came with his force. The incident took an ugly shape with firing and violence. A Nishi got killed in this incident. Some time passed and in November 1992, the Nishis took revenge by attacking a bus full of Apatanis passing by their area. The bus was burnt down and the conflict took the shape of communal violence. Thereafter the Nishis reinforced the boundaries of their forest and started patrolling the area. Richop fled from the State for sometime. At such a time, the Central Home Minister had to intervene and the administration under the chairmanship of the DC was asked to form a committee and deal with the situation. Nothing much happened on this front but on an appeal from the Chief Minister and Home Minister of the state, the dispute subsided but it has never been resolved.

At this time, the local village council decided to take up the matter as it had flared up beyond a limit. The council decided to investigate and administer justice in this case under their customary laws. They tried to find out the person responsible for instigating communal violence in the society which resulted in a considerable amount of destruction. A series of meetings were conducted by the elders of the Apatani tribe at different *lapangs* in a series on 18th, 19th and 20th April 1993 to find out the cause of the communal clash between the two tribes. In these meetings it was decided to seek clarification from the two Apatani youth leaders Padi Richop and Nani Ribya, whom people of both the communities had blamed for creating unlawful problems in the district and the *buliang* decided to penalise both the leaders with a deposit of Rs. 50,000 (US$ 1250) each. Padi Richop deposited the money on 22nd April but Nani Ribya did not do so despite many reminders. To decide upon the matter, the village elders of three villages met on 28th April, where the *buliang* decided to invite Padi Richop and Nani Ribya to perform a *kashu* (chanting of omens) and the date for the next meeting was fixed for 30th April. In the meantime, on 29th April, Nani Ribya gathered his well-wishers in his village and declared himself innocent of the communal riots. He boycotted the 30th April meeting and instead held a community feast in his village for his kinsmen. Subsequently, on 2nd May, after scrutinising all evidence, the *buliang* declared Richop innocent and held Ribya responsible for the riots. The *buliang* on having pronounced its judgement left the matter with the District Administration for a prompt and stern action to be taken against Nani Ribya and to hold him responsible for all the loss to human life and damage to property. A victory celebration was held at Ziro on 11th May, where three Mithuns, four cows and other domestic animals were slaughtered by Richop and the function was attended by nearly 2000 people.[33]

Note: Case number seven and eight reflect that cases, when taken to formal State judicial bodies, do not get resolved; either such cases languish in the courts for years without being heard or justice met, or they are sent back to the State executive bodies for disposal who again are not able to come to a decision in critical matters.

---

33. This case was reported by Bengia Aubung (a Nishi journalist) in a news magazine, the *Eastern Panorama*, in June 1993 and is entitled 'Tribal Panchayat Holds Congress Leader Responsible' (Arunachal Communal Clash).

9. A case which has had far reaching implications not only on the forests of Arunachal Pradesh, but of the entire country is the case of T.N. Godavarman Thirumulkpad and others versus the Union of India and others.[34] The first interim order of the Supreme Court on 12th December 1996 in this case granted a blanket ban on felling of timber in all forests and directed the application of the Forest (Conservation) Act, 1980 to all forests irrespective of ownership or classification. In some areas, the court directed a ban on felling of any kind of tree. The court even issued a ban on the movement of cut trees and movement of timber from the north-eastern States of India. Functioning of all saw and veneer mills was halted in some parts of Arunachal Pradesh as per the order. Implementation of such an order in the State of Arunachal Pradesh is a near impossible situation. Considering the vast stretches of remote forests in the State and the limited resources of the Forest Department, monitoning the implementation of such an order is impractical. Moreover, enforcement of provisions of the Forest Conservation Act, 1980 to the State seems a difficult task, as over 80 per cent (unofficial figure) of the State is under forest cover and to seek the permission of the Central Government to clear any forest land for the pursuance of any non-forest activity also seems a bit unworkable, especially in this case where most of the forests are under community management. This case shows the mindlessness of the formal judicial body in decreeing such an order. As against this, empowerment of the traditional institutions in this regard with some kind of a monitoring element placed thereupon by the state could serve as a more feasible solution to the problem.

---

34. Writ Petition (Civil) No.202 of 1995 including WP 171/1996, Supreme Court.

Recent developments

Arunachal Pradesh is at the juncture of transformation of economic, social and political structures and tribal value-systems. Hitherto, the effectiveness of tribal laws, tribal conflict resolution mechanisms, tribal concepts and systems of justice and the indigenous institutions cannot be undermined in the realm of forest resource management. It is to be seen in the future as to how far and how long these systems will sustain and continue to be effective, though the reason for these systems being effective so far is the inherent element of change and adaptation involved therein. Customary laws differ from statutory laws in their efficacy mainly due to the reason that the former are flexible and adapt themselves to the changes in society; whereas to amend a statutory norm requires immense political will and it is a time-consuming process. The Indian Forest Act, 1927 is a case in point. This piece of legislation was a colonial imposition and enacted by the British Government with the main objective of exploitation of the forest resource for economic reasons. Even 50 years after Independence, the Indian Parliament has not been able to revoke the law and enact a new one in its place to meet with the requirements of the new situation.

The Forest Policy of 1952 for the first time acknowledged the ecological aspect of the forests and tacitly affirmed the needs of the people living from the fringe benefits of the forests by referring to the need to establish tree-lands and small woods for grazing and firewood. The Forest Policy of 1988 has explicitly recognised peoples' needs and role in conservation.

The new situation that necessitates conservation of biological resources is reflected in the Forest Conservation Act, 1980 and some progressive policies formulated by the Indian Government. The more recent approach to forest management, Joint Forest Management of 1990 is now also being promoted by the government. These policies merely serve as guidelines and the provisions of policies are not enforceable in the court, hence become meaningless for a concerned citizen. The JFM circular sets out a new policy of involving village communities and voluntary agencies in the regeneration of degraded forest lands. This has been introduced in Arunachal Pradesh in 1997 and has been intended to extend to the degraded USFs. The implementation of this could lead to resentment in the field as the local people do not have any faith in the Forest Department. In addition, inter-tribe boundary conflicts exist in the society, which may hinder the smooth implementation of the JFM scheme in the State.

Since Arunachal Pradesh was exempted from the application of the centrally legislated Indian Forest Act, forest management had largely been the prerogative of the local communities. It was only in 1916 that the Assam Forest Regulation of 1891 was extended to Arunachal. The Assam Forest Regulation is drafted on the lines of the Indian Forest Act. Although the provisions of the two laws are similar, the enforcement of the law in the State due to many factors had no affect upon the control of the local communities on the forests. The local people continue to regard the forests as theirs. Some Reserved Forests were notified in Arunachal Pradesh in the past but with the growing awareness among the local people regarding the value of the forest resources and also the provisions of law in relation to the procedure of declaration of Reserved Forests, many instances of challenging these creations in the court are coming to light.

The Forest Conservation Act, 1980 is the first enactment in the post-Independence era depicting some conservation goals. The Forest Conservation Act envisages restrictions on the use of forest land for non-forest purposes. It seeks the prior approval of the Central Government for the purpose, thereby centralising the whole process. This seems a bit impractical for the State of Arunachal Pradesh. Considering the liberal definition accorded to the term forest under this Act, most of Arunachal would fall under the category and the Centre would be flooded with applications for use of forest land for different purposes.

The government has tried half-heartedly to undo the effects of the forest legislation, but this has had a negligible impact on the forests of Arunachal Pradesh. The only effective effort so far in stemming the illegal felling of timber is that made by the Supreme Court of India which in December 1996 had banned the felling of timber and movement of timber from this State. Although this order of the Supreme Court has been effective in curbing the illegal extraction of forest resources, it has led to social disorder of another kind. This has taken away the only source of income from the people and the revenue earned by the State has decreased thus leading to reduced government investments. This order has left many people unemployed.

There is a growing consensus that law inherited from colonial times is inappropriate to regulate use and management of forest resources. Whatever the limitations of the customary or community framed laws are, they are still effective in governing local management of forest resources. This also holds true for traditional institutions, which have the capacity to achieve sound management through effective conflict resolution. For the purpose, it becomes imperative to rely on local laws duly legitimised by the state. Mere policy formulations do not affect forest management unless these are translated into legislative actions.

References

Cathie, J. Willie, 1996, "Comparative Jurisprudence" in: Hunter and Whitter (eds.), *Encyclopedia of Anthropology*, New York, Harper Row, pp. 84-5.

Das, J. N., 1986, *A Study of Administration of Justice Among the Tribes and Races of North-Eastern Region*, Law Research Institute, Guwahati High Court, Guwahati, pp. 312-26.

Fürer-Haimendorf, Chr. von, 1980, *A Himalayan Tribe: From Cattle to Cash*, Shahibabad: Vikas Publ. House.

Government of Arunachal Pradesh, n.d., *Arunachal Pradesh Forest Statistics 1990*, Forest Department, Itanagar.

Government of Arunachal Pradesh, n.d., *Arunachal Forests 1993*, Department of Environment and Forests, Itanagar.

Moore, S.F. 1973, Law and Social Change: the Semi-Autonomous Social

Field as an Appropriate Subject of Study, *Law & Society Review*, vol. 7 (4):719-47.

Myers, N., 1988, Threatened biotas: 'hot spots' in tropical forests, *The Environmentalist*, 8 (3):187-208.

Patnaik, S.S., Barik, S.K. and Shukla, G.P. 'Arunachal Pradesh Anchal Forest Reserve Act 1975, Village Forest Reserve Act 1984 and New JFM Resolution: A Conceptual Change from Partnership Management of High Forests to Participatory Forest Management', paper presented at the JFM workshop in Itanagar on 3 and 4 October, 1997.

8

Indigenous Knowledge in Traditional Modes of Forest and Tree Management in the Central Himalaya, India

N.P. Melkania

Resource use in the Central Himalayas

THE Himalayan mountainous forests have been acknowledged as resources overriding the regional and national environmental economy of India and adjacent border countries, particularly Nepal and Bhutan. This environmental economy is evident both in the form of intangible and tangible benefits, such as the regulation of atmospheric gaseous exchange, control of soil loss, availability of perennial water bodies for drinking water and various aquatic resources, great biodiversity in flora and fauna, control of floods downstream, timber and other industrial raw materials and various non-timber produces. Forest degradation has been serious in the twentieth century and the environmental repercussions of deforestation are massive and evidently observed as loss of biodiversity, scarcity of forest-based raw materials, decline in soil productivity, drying-up of springs and drinking water resources, siltation and floods.[1]

Comprising of eight hill districts of Kumaun and Garhwal divisions of the state Uttar Pradesh, the Indian Central Himalaya (*fig.*1) experiences explosive biotic pressure on the natural forests

1. Melkania *et al*. 1988.

for the national and local economic needs. The region is inhabited by people of a wide range of ethnic composition. The caste groups like Rajput, Brahman, Harijan (Shilpkar or scheduled caste) and Janajati (scheduled tribe) like Bhotia and Raji dwell in close proximity to each other in this region. There are many similarities in the way of life among these people, but they exhibit certain specific differences in their life and their living patterns, too. The tribals as defined politically, viz., Bhotias and Rajis have been leading a semi-nomadic and transhumant life because of historic and environmental factors which was till recently associated with the Trans-Himalayan trade of the Bhotias and food-gathering and hunting activities of the Rajis.[2]

The major activities of the 'ecosystem people' (people solely dependent on their own ecosystem) in relation to forests in the Central Himalaya are as follows:

- Nomadism with pastoral activities, e.g., Gujjars.
- Semi-nomadism with semi-pastoral activities, e.g., Bhotias.
- Settled farmers with shifting agriculture (e.g., farmers in parts of Pindar Valley in Kumaun) and settled agricultural activities.
- Industrial work combined with tea and fruit plantations, e.g., urban farmers.

The natural vegetation in this hilly region represents sub-tropical semi-evergreen Sal (*Shorea robusta*) forests in the foothill; coniferous forests dominated by Chir-Pine (*Pinus roxburghii*) and Deodar (*Cedrus deodara*), mixed Chir-Pine broad leaf forests and pure broad leaf forests of Abies and Picea and high altitude scrub lands in the high mountain regions.[3] Legally, the forests are categorised into government forest (reserved and protected)

---

2. N.P. Melkania, 1983; N.P. Melkania and Melkania, 1988; Kapoor *et al.*, 1990.
3. Uma Melkania and Melkania, 1988.

and community-owned (civil/soyam and Panchayati) forests. The former are managed by the State Government Forest Department and the latter are administered by the State Revenue Department. The government forests occupy 45.76 per cent of the total area (51943 km^2) of the region. The civil/soyam and Panchayati forests occupy, respectively, 14.63 per cent and 4.72 per cent of the area. At the regional level among the Central Himalayan districts, the forest area per hectare of the reporting area in 1983-84 was only 0.64 ha which ranged from 0.52 ha in Almora to 0.89 ha in Uttarkashi. The total forest area in the region in 1983-84 was 67.13 per cent of the reporting area (53,35,065 ha).[4] Studies based on satellite imageries indicated that only 28.7 per cent of the officially claimed total forest area (14,638 km^2) is actually under forest cover; the good forest cover (more than 60 per cent crown cover) is accounted only for 4.4 per cent (2263 km^2) of the total forest area (Table 1).[5]

Table 1: Forest area in the Indian Central Himalaya

Categories	*Forest area (km^2)*
Poor Forest	4709
Medium Forest	7666
Good Forest	2263
Total Forest	14638

Poor forest: Crown Density (C.D.) < 30 per cent, Medium Forest: C.D. 30-60 per cent and Good Forest: C.D. > 60 per cent

The people's forests

Of the forests categories described above, truly speaking, Panchayati forests (syn. Van Panchayat forests, Gaon or Gram Panchayat forests) represent the peoples' forests (syn. community forests) in Central Himalaya. Unlike the civil/soyam forests (protected forests) where the reforestation and other rehabilitation

4. N.P. Melkania and Melkania, 1988.

5. Singh *et al.*, 1984.

activities are accomplished by the State Forest Department after an agreement with the concerned villagers and revenue authorities, the Panchayati forests are strictly under the control of the Revenue Department. Even the reforestation and rehabilitation activities are controlled and regulated by the district administration through the forest Panchayat inspector (earlier) and recently the district afforestation officer and the forest Panchayat inspector collectively with the technical assistance of the State Forest Department officials. The village-level activities are regulated and monitored by the Van Panchayat headed by the Sarpanch. In some villages, the Panchayati forests are watched by a guard (local villager). However, keeping a guard is not an administrative necessity. The State Revenue Department administers the Panchayati forests by Van Panchayat Niyamawali.

People's perceptions of forests

The Central Himalayan people in general identify forest as a plant community where trees constitute the dominant plant forms and determine their survival in the rugged mountainous environment. The inhabitants (rural people) define local use of the forests as the support for their major economic activities; for agriculture, composts, farm implements, bedding material and fodder for their animals; as the only never ending and locally available source of wood energy for cooking and water and space heating; as a source of timber for constructing houses and small furniture as needed; as a source of natural rains and water; and as a source of local enterprise, such as firewood selling, grass selling and collection of wild edibles. People are also aware of the environmental role of the forests in regulating natural spring recharge, as a control of soil loss, as an abode for wild animals and as a regulator of floods. The term 'local use' being used in the paper reflects the people's concept of forests.

The people's management of community forests

The Panchayati forests belong to the land-based common property resource category (CPR) managed by the 'ecosystem people' for their local uses. These resources earlier were managed traditionally, prominently on the basis of social values. The position of the Sarpanch was sometimes inherited by the generations where the social values were more than the position/ administrative status; otherwise the new Sarpanch was elected by common consensus. The process was also similar for electing the committee members. The committee headed by the Sarpanch was authorised for administering decision-making and outside contacts concerning the Panchayati forests. The committee collectively or the Sarpanch individually used to seek matters, as and when needed. The watch and ward duty used to be done by the villagers on rotation-basis and through a guard identified by the committee. In the latter case, the guard used to collect a fixed quantity of agricultural produce at the end of each crop season as his wage; the payment in cash developed later.

The vegetational resource of the Panchayati forests used to be developed without any imported technology by the villagers themselves. The first step for the development of such forests used to be the effective protection through fencing from disturbances by domestic cattle (grazing) and anthropogenic activities (burning, cutting of trees, lopping, debarking, litter removal and non-wood forest produce collection). The villagers even used to leave dry and dead materials on the forest ground for natural recovery to recharge the ecological processes. The vegetal cover used to be enriched by transplanting naturally developed seedlings from the adjacent forest areas and privately managed grasslands (Ghasnis). In dense regeneration stands, the tree crops used to be thinned to facilitate tree growth for optimum productivity. The diseased and malformed tree saplings used to be uprooted or harvested.

Grass fodder, wood for cooking and timber for house construction used to be the principal tangible benefits from these resources and were shared with equity (such as one sickle per family at one time for grass cutting). The committee members, the Sarpanch and the concerned villagers used to consider community resource protection as their moral duties because of high respect for social values and cultural norms. The Panchayati forests, thus, used to be managed as CPR in an integrated management system considering ecological, technological, economic, ethnological and institutional aspects simultaneously and sequentially, where sustained supply of forest resources used to be the criteria of forest health.

Post-Independence scenario

There has been a profound change in the human values in the mountain villages particularly after Independence because of increased political awareness and urbanisation and modernisation. People now prefer to live in urban centres or in villages with urban facilities. They prefer public and private sector services to the village-level local service, as the former services are valued higher in society. The social values are gradually replaced by economic-political values and urban elitism. Activities like cultivation of trees, agriculture, etc., have become areas of concern for uneducated, undereducated and old or retired people. Emigration of educated village youths in search of a job in nearby towns and plains also accelerated the change in the value system.

This change has also influenced the management of Panchayati forests. Now the respect for the conservation of natural and community resources has declined substantially even among the committee members and the Sarpanch. The appointment of a guard for watch and ward has now become a necessity. The administration and management of the Panchayati forests many times become part of the village-level politics. The earlier referred to symbiotic relationship of the Sarpanch, the committee members and the villagers is evident only in villages where social values are still respected and demanded. This change in human values has led to the degradation of the Panchayati forests in the majority of places much more than the local population pressure. Now dense and diverse Panchayati forests are available only in areas where government forests exist adjacent to villages (as the people collect forest produce from the government forests) or where the Sarpanch still commands respect.

The state government, consequently, decided to support Van Panchayat exogenously with technical inputs much more than in earlier days. The attempts made by the administration in the Almora district are worth mentioning. During 1975-77, the then district magistrate had been instrumental in developing a Van Panchayat movement in the district in particular, and in the Kumaun region in general. Under his acknowledged leadership and guidance the revenue authorities made serious attempts for the regeneration of the dying Van Panchayat and establishing Van Panchayat in villages where such systems were absent. Though the people were motivated for raising Panchayati forests, the objectives in a practical sense were yet to be achieved. The major problems in this regard have been the Van Panchayat rules conflicting with the participatory approach, a change in the value system as referred to earlier and a lack of regular efforts for making such programmes a people's programme in reality.

In the early years of the Green Revolution, the slogan 'grow more food' also accelerated the loss of trees and land under tree cover, as community land, degraded forest land and marginal rain-fed land areas were also influenced. The density of trees in the farm land was reduced drastically, prominently in the valley agriculture to increase agronomic crop cultivation. Land which was otherwise unsuitable for raising agronomic crops was put under the plough. Further, the government policy concerning providing land to the landless also reduced the area under tree cover and accelerated soil erosion with low agronomic yields from these sites. The scientific research concerning a better land use is so scanty that the research on agroforestry could not view beyond the outer foothill region of the Duns, Bhabar and Tarai.

Tree cultivation strategies in croplands

In the Indian Central Himalaya, permanent cultivation is practised only up to 2700 m, while in areas between 2700-4200 m farmers cultivate crops only during May to September.[6] No trees are

---

6. N.P. Melkania, 1983.

grown by the farmers on cropland above 2700 m. In the *Outer Himalayan Region* (300-700 m a.s.l., sub-tropical) the farmers practise intensive agriculture and cultivate some trees on farmlands. Most of the farmland trees are fruit-yielding while trees of other domestic uses are cultivated in wastelands, roadside lands and other barren uncultivable areas (Table 2). Recently, the farmers in the Tarai region have started the cultivation of industrial tree species, predominantly *Eucalyptus tereticormis* and *Populus deltoides* G-3 and G-48 under farm forestry and agroforestry.[7] But most of the traditional farmers still prefer conventional tree species of domestic importance.

The farmers living in the *Middle Himalayan Region* (700-2300 m a.s.l., predominantly temperate) generally practise terrace cultivation for complete subsistence or, in addition, some cash crops, e.g., potatoes, onions and other vegetables; fruits like apples, pears, plums, etc., to satisfy their families' needs for fuel, food and marketable products. Most of the agronomic area is rain-fed or upland; irrigation is possible only in lowland valley areas. Tree cultivation on the terrace risers within the farm land (traditional agroforestry) is often a first step toward personal, economic and ecological stability. But such practices are rare under valley cultivation conditions. The unapproachable and/or less productive agronomic fields are left fallow temporarily or permanently to let natural grasses grow. Thus, these landscapes act as privately owned grazing lands without or with woody shrubs and tree saplings. The latter are managed as savannah or silvipasture.[8]

Most of the trees cultivated on farmlands in this region are non-forest species, predominantly cultivated for top feeds, farm fences (e.g., *Rubus ellipticus*, *Prinsepia utilis* and *Berberis* spp.) and for other domestic uses, e.g., fruits, fibre, *patta* (leaf plate) for local festivals and wooden utensils (e.g. *Boehmeria rugulosa*) (Table 2). Tree species, viz. *Pyrus pashia* and *Sapium sebiferum*

---

7. Prasad, 1987.

8. N.P. Melkania and Tandon, 1985.

Table 2: Tree species cultivated by local farmers in the Central Himalaya

Species	Local name	Principal use	OCZ			MCZ			ICZ		
			AF	AbF/ GL	OA	AF	AbF/ GL	OA	AF	AbF/ GL	OA
Acacia catechu	Khair, Kattha	Kattha		+	+						
Azadirachta indica	Neem	M		+	+	+					
Albizzia Lebbek	Siris	Fi Fo		+	+	+					
A. procera	Siris	Fi Fo		+	+	+					
Alnus nepalensis	Udis	Fi					+	+		+	+
Artocarpus integrifolia	Kathal	Fr		+							
Boehmeria rugulosa	Gethi	Fo Wu				+					
Bauhinia purpurea	Qaueral	Fo F1				+					
B.retusa	Kanol	Fo				+	+	+			
B. vahlii	Malu	Fo Patta		+	+						
B. variegata	Kachnar	Fo F1	+								
Berberis asiatica	Kilmora	Fo M F				+	+	+	+	+	+
B. Lycium	Kilmora	Fo M F				+	+	+			
Bombax ceiba	Semal	Fi C	+	+	+	+	+	+			

Species	Local name	Principal use	OCZ			MCZ			ICZ		
			AF	AbF/ GL	OA	AF	AbF/ GL	OA	AF	AbF/ GL	OA
Carica papaya	Papita	Fr	+								
Carpinus viminea	Chamkharik	Fo Fi							+	+	+
Celtis australis	Kharik	Fo Fi				+	+				
Citrus spp.	Nibu	Fr	+			+			+		
Dalbergia sissoo	Shisham	T Fi	+	+	+						
Diplochenma butyracea	Chiura	Fr Fo		+		+	+				
Emblica officinalis	Amla	Fr	+	+	+	+	+	+			
Erythrina suberosa		Fi F		+	+		+	+			
Eugenea jambolana	Jamun	Fr		+	+	+					
Ficus palmata	Beru	Fo Fr				+	+				
F. religiosa	Pipal	Fo R		+	+		+	+			
F. roxburghii	Bargad	R			+			+			
Ficus spp.		Fo							+	+	+
Grewia optiva	Bhimal	Fo Fi Fib				+	+				
Jacaranda spp.		Fi S O			+			+			

Species	Local name	Principal use	OCZ			MCZ			ICZ		
			AF	AbF/ GL	OA	AF	AbF/ GL	OA	AF	AbF/ GL	OA
Juglans regia	Akhrot	Fr				+	+	+			
Mangifera indica	Aam	Fr	+			+					
Melia azedarach	Baka in	Fo Fi			+	+	+				
Morus alba	Sahtoot	Fo Fr				+					
M. serrata	Kimu	Fo Fr				+					
Musa spp.	Kela	Fr Fo	+			+					
Pinus roxburghii	Chir-pine	Fi T Re					+	+			
Prinsepia utilis	Jhitalu	F O				+	+				
Prunus cerasoides	Paiya	Fo R				+	+				
Psidium guajava	Amrud	Fr	+			+					
Pyrus communis	Naspati	Fr				+					
P. malus	Seb	Fr				+			+		
P. pashia	Mahal	M				+	+	+			
Pyrus spp.		Fr				+					
Quercus floribunda	Tilaunj/ Moru oak	M				+	+	+			

Species	Local name	Principal use	OCZ			MCZ			ICZ		
			AF	AbF/ GL	OA	AF	AbF/ GL	OA	AF	AbF/ GL	OA
Q. leucotrichophora	Banj oak	M				+	+	+			
Q. semecarpifolia	Kharsu oak	M							+	+	+
Rubus ellipticus	Peela Hisalu	F Fr Fo				+	+	+			
Salix alba		Sc Fo Fi							+		
S. babylonica		Sc Fi						+			
Sapindus mukorossi	Ritha	Soap				+	+	+			
Sapium insigne	Khin	F				+	+				
S. sebiferum	Charbi	Fi Fs				+	+				
Terminalia arjuna	Arjun	S			+			+			
T. belerica	Bahera	Fr			+		+	+			
T. tomentosa	Harra	Fr			+	+	+	+			
Toona ciliata	Tuni	T Fo				+	+				
T. serrata	Suni	T				+	+				
Vitex negundo		F Fi				+	+	+			

Abbreviations used in the Table 2:

AbF/GL = Abandoned Fields/ Grazing Land
AF = Agricultural Fields
C = Cotton
F = Fence
F1 = Flower
Fi = Firewood
Fib = Fibre
Fo = Fodder
Fr = Fruit
Fs = Fodder Storage
ICZ = Inner Cultivation Zone
M = Multipurpose
MCZ = Middle Cultivation Zone
O = Ornamental
OA = Other Areas
OCZ = Outer Cultivation Zone
Patta = Leaf plate
R = Religious
Re = Resin
S = Shade
Sc = Soil Conservation
T = Timber
Wu = Wooden Utensils
\+ indicates availability of the species

are cultivated for open storage of hay and straw; the former is also cultivated for transplanting pear grafts on it. Such tree cultivation strategies contribute to people's farming efforts by reducing fencing and storage costs, providing needed shade for perennial cash crops and adding forage for cattle. Most of the tree species are persistent coppicers and widely recognised by the small farmers for their restoration characteristics as nitrogen fixers improving the quality of the soil. Their foliages decompose faster than most of the forest species. Species, viz., *Toona ciliata*, *T. serrata* and *Quercus* spp. constitute a real investment for future generations when small farmers protect it.

The *Inner Himalayan Region* (2300-2700 m a.s.l.) is located close to sub-alpine zone where tree cultivation suffers due to the vagaries of the weather. Further, compared to the mid and outer Himalayan regions, this zone is less populated and represents rich and luxuriant wild vegetation. Hence, the cultivation of trees in this region is less common. The farmers gather their domestic tree-based goods from adjacent forests. They practice subsistence farming and in home gardens the cultivation of apples, pears, cauli crops and potatoes is common. In some villages of the Pindar valley in Kumaun, e.g., the villages around Khati and Dwali, farmers practice shifting cultivation in bush invaded areas. The tree species cultivated by the farmers of this region (Table 2) are mostly the forest species as they are the only known suitable woody species for this climatic region to be used in agriculture.

Lessons to be learned

In the Central Himalayan Region, where land holdings are small, fragmented and scattered at several locations in the villages and outside, tree planting on farmland is rarely a novel activity for subsistent farmers. It is often part of their farming tradition, ever since many generations. In fact, the farmers have developed traditional tree cultivation strategies in response to locality and climatic factors and their basic needs, as alternative food sources, as a source of cash income, where market demand for the wood and other tree products exists (e.g., timber), as substitutes for purchased products (e.g., firewood, fibre, fence posts, construction material, etc.) and as labour-saving or labour-spreading techniques (e.g., living fence posts). These strategies are characterised by the following features:

- The inputs are strictly indigenous, simple, dependent on the farmer's participation during slack labour periods. The propagating materials are available for the farmers.
- Farmers tend to prefer species that reproduce vegetatively from cuttings, root stumps or root stocks, suckers or coppicing.
- Many tree species, particularly those used for plantation in agricultural fields or as shade trees, tend to be multipurpose.
- Most of the tree species planted in farm lands are deep rooted, hence, there is negligible competition between tree species and agronomic crops for nutrients and moisture availability.
- The foliages of most of the farmland tree species are smooth and soft and tend to decompose easily and quickly to buildup organic matter to fertilise the soil.
- Planting fodder trees in farmland (agroforestry) and privately owned grazing lands (silvipasture) overcomes the problem of scarcity of land for fodder cultivation and green fodder during the lean period.
- Planting trees in agricultural systems reduces fencing costs, provides needed shade for perennial cash crops and adds forage for domestic cattle.
- Planting diverse species of fruit and other economic trees helps spreading the risks involved with the exclusive dependence on annual root or grain crops, grass fodder and single top feed species, which may fail owing to climatic disruptions or due to insect-pests.

- Most tree species are tolerant to cleaning, pruning and other tending operations. Most tree species can wait for maintenance until the farmer has slack time or when the operations coincide with another cultural practice linked to the associated crops. Most trees can tolerate the delayed cleaning without being lost completely.
- Small landholders may invest more time of the non-peak labour period for planting and maintaining industrial and other commercial tree species, even when they know it will be their children who will benefit (e.g., *Toona ciliata* and *T. serrata*).

Reflections and conclusion

The foregoing description evidently indicates that ligneous elements, specially tree species, are planted by the 'ecosystem people' in the Indian Central Himalaya with crops on farmland, margins of the agricultural fields, abandoned lands and natural landscape areas, such as community lands and natural forests. The strategy for managing the tree cover varies in different habitats, such as crop lands and natural forests. It is evident that these vegetational components are managed by the 'ecosystem people' in traditional modes by the use of indigenous knowledge, which is based primarily on local ecological and social conditions. Such strengths can be effectively used in managing trees and forests in this region on a sustainable basis.

The management of natural forest is a complex affair which demands the understanding of the natural forest dynamics and takes into account human considerations, so that the actors and factors for developing management strategies can be identified appropriately and used judiciously in management planning. The management policy must be motivational and consider three important aspects, viz., the livelihood of the 'ecosystem people', the quality of the environment both bio-physical and socio-economic and the genuine proprietorship of the resources. The following points need to be noted with emphasis on the management of forest in the Himalayan region:

- An effective management of natural forests can be achieved by giving focussed value for those who live with them.
- Differential inputs must result in differential benefits. The policy must ensure that the benefits are strictly related to the inputs.
- There must be a positive correlation between the quality of management and the benefits.
- The unit of proprietorship should be the unit of production, management and benefit. The unit of proprietorship should be as small as practical within the ecological and socio-political considerations.

As indicated, the forests in the Central Himalaya belong to the State Forest Department, Revenue Department and local people in categories, viz., reserve, civil and soyam and village forests, respectively. These categories may be managed in the following ways:

- The reserve forest areas can be managed by the State for regulatory functions and occasional productive functions as sites of natural and ecological significance.
- The degraded forest land in the reserve forest area may be collectively managed and developed effectively by the State Forest Department, forest dependent industries, NGOs and local people. In this agreement, the Forest Department can provide the technological know-how, the industries should provide the financial support and the NGOs can act as interface. The local people can be used as human resource. The users, viz., industry and local people may be provided with raw material and wages, respectively, for their contribution.
- The community-owned forests or village forests can be managed effectively by integrating the strengths of the so-called Joint Forest Management in the Van Panchayat

system. Here, the State Forest Department should impart the technological know-how, the NGOs can act as interface and the local people as partners and human resource. Instead of providing wages to the local people, usufruct sharing may be adopted as an approach to reward the contribution of the local people. This programme would require a truly participatory approach focussing primarily on the 'ecosystem people' and their interlinked ecological and socio-economic conditions. A suitable reward and punishment system and post project maintenance activities need to be integrated in this approach.

As indicated in the foregoing pages, the traditional management of community forests in the region illustrates an example of integrated resource management. However, with the change in time and human values, this needs to be strengthened further. The technological aspects need to be integrated with the ethnological and economic conditions of the local people so that an ethno-economically sound technology is developed for managing productive and diverse community forests. The ecological aspects need to be integrated with the policy so that it is not overridden by the economics. Possibly, an understanding of ecological economics may solve this conflict of development and management. The village institutions need to be strengthened considering the non-elite group of the villagers, ethnic categories and gender. Care must be taken to identify the representation and deciding majority in decision-making and usufruct sharing, so that silent and weaker sections are protected effectively from the power of elite villagers. An attempt has been made in this regard by the author in an earlier publication.[9]

With regard to developing the tree component on farmland, adequate emphasis should be given to the characteristics of hill agriculture, such as small and fragmented land holdings, rain-fed and marginal cultivation, subsistence economy and unique

---

9. N.P. Melkania, 1993.

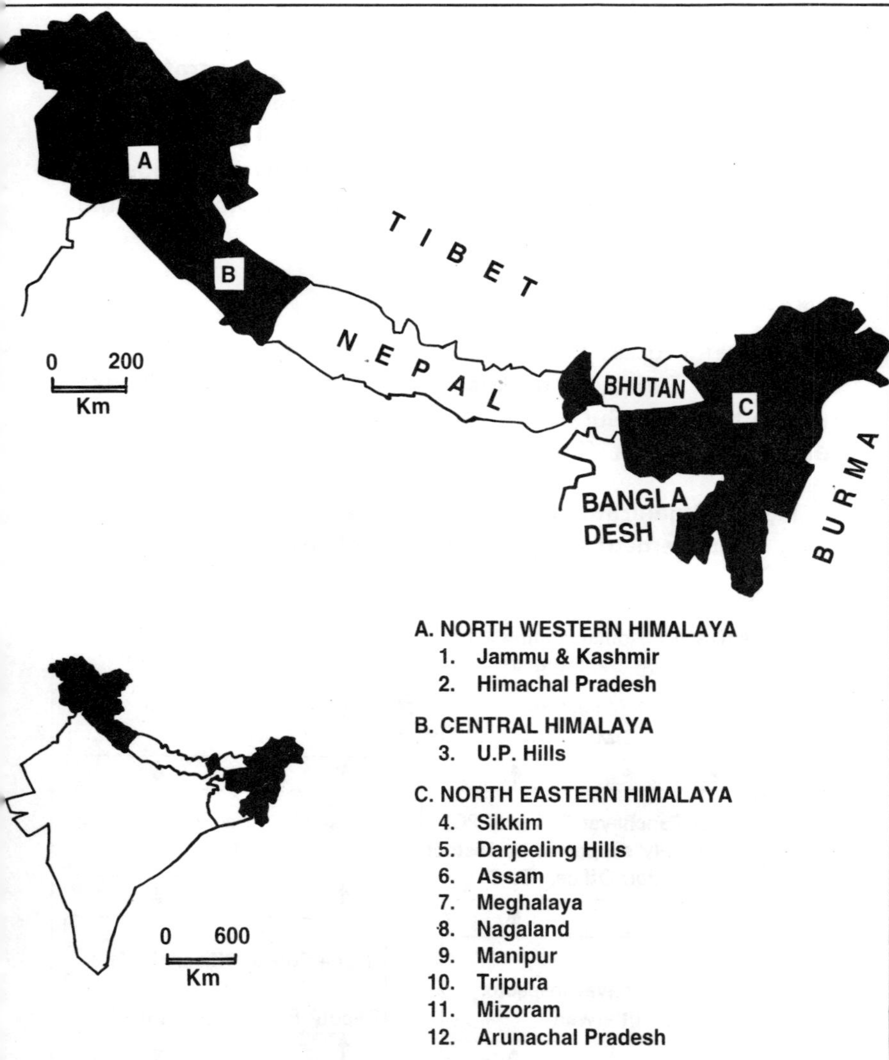

Fig. 1: The Himalayan Range

cultivars specific to the region, such as horse gram (*Cicer aerietinum*), bhat (*Glycine soja*) and pseudo-cereals. Recalling that the cultivation is practised on terraces, the agroforestry models of the foot hills and plains must not be transplanted to the region. Probably technological developments like agroforestry, horticulture and eco-development need a rethinking considering the interlinked ecological and socio-economic factors of the region.

Acknowledgements

I am grateful to Dr. Klaus Seeland, Swiss Federal Institute of Technology, Zurich, for his keen interest in the subject. Thanks are due to Mr. S. Suresh Kumar and Mr. Shekh Aslam for able secretarial support. The financial support from the organisers

Fig. 2: Organizational structure of the Panchayati forest management system in the hills of Uttar Pradesh

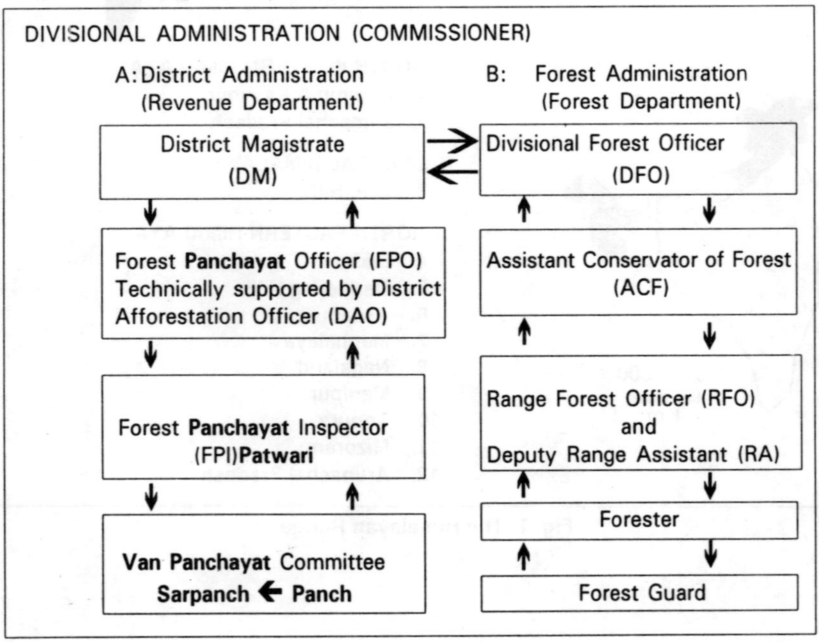

and the administrative support from Shri B.P. Singh, Director of the Indian Institute of Forest Management, Bhopal, for my participation in the National Seminar on Indigenous Knowledge and Forests are also acknowledged gratefully.

References

Melkania, N.P., 1983: Report on the surveys and germplasm explorations conducted during 1982 in the Indo-Tibet border regions (tribal areas) of Kumaun Himalaya, Vivekananda Laboratory for Hill Agriculture, Almora.

Melkania, N.P., 1993: Developing management of the forests for the local use in the Himalayan mountains, In: Rawat, M.S.S. (ed.): *Himalaya: A Regional Perspective (Resources, Environment and Development*), Daya Publishing House, New Delhi, pp. 99-106.

Melkania, N.P. and Tandon, J.P., 1985: Dry matter yield and strategies for regeneration of pastures in central Himalaya, In: Singh, J.S. (ed.): *Environmental Regeneration in Himalaya: Concepts and Strategies*, CHEA and Gyanodaya Prakashan, Nainital, pp. 399-409.

Melkania, Uma & Melkania, N.P., 1988: Forest and Forestry in Uttarkhand, *Uttarkhand*, 2, pp. 97-138.

Melkania, Uma & Melkania , N.P. 1988: The Human Environment in Himalaya, *Man in India*, 68, pp. 121-44.

Melkania, Uma, Melkania, N.P. & Jhori, B.N., 1988: The state of forests in Kumaun and their regeneration, In: Valdiya, K.S. (ed.): *Kumaun – Land and People*, Gyanodaya Prakashan, Nainital, pp. 243-51.

Prasad, S., 1987: Agroforestry patterns in Tarai belt of Nainital district and influence of leaf leachates and litter of Eucalyptus and Populus on some growth parameters of wheat and gram. M.Sc. Thesis, G.B. Pant. University of Agri. and Tech., Pantnagar.

Singh, J.S.; Uma Pandey & Tiwari, A.K., 1984: Man and Forests: A Central Himalayan Case Study, *Ambio*, 13, pp. 80-7.

9

Natural Resource Management and the Role of Local Knowledge in the Western Himalayas

Sudha Vasan

> Indigenous people and their communities represent a significant percentage of the global population. They have developed over many generations a holistic traditional scientific knowledge of their lands, natural resources and environment.... In view of the inter-relationship between the natural environment and its sustainable development and the cultural, social, economic and physical well-being of indigenous people, national and international efforts to implement environmentally sound and sustainable development should recognise, accommodate, promote and strengthen the role of indigenous people and their communities.
>
> — Agenda 21, Chapter 26
> United Nations Conference on
> Environment and Development
> Rio de Janeiro, 1992

Introduction

THE last two decades have seen increasing recognition of the knowledge of local, tribal, or indigenous people as valid, legitimate, useful and possibly the panacea to managing most environmental problems. This has emerged as a retort to the centralising and universalising development theories of the fifties and sixties that trivialised indigenous knowledge and characterised it as an obstacle to progress. The derogatory imagery of local knowledge as being inferior to Western science has been somewhat cleared by the many empirical studies from around the world. The first systematic studies on indigenous knowledge started in the late seventies (Howes *et al.* 1979, Brokensha *et al.* 1980) but it is the eighties that saw the resurgence and deification of this field. Initially it was only anthropologists or development sociologists who were interested in this field, but now

> ...important contributions are being made in the fields of ecology, soil science, veterinary medicine, forestry, human health, aquatic science, management, botany, zoology, agronomy, agricultural economics, rural sociology, mathematics ...fisheries, range management, information science, wildlife management, and water resource management.
>
> — Warren et al. 1993:2.

This has resulted in varied understandings of indigenous knowledge and varied ideas about its utility.

The concept of indigenous knowledge has, however, completed the whole circle. A number of studies have emerged that have taken a more critical look at the concept of indigenous knowledge (Agrawal 1995). The concept of indigenous knowledge, particularly the forms it has taken in Western academia, has created a tension and a false dichotomy between Western scientific knowledge and indigenous knowledge. The romanticisation of the concept has also led to claims about its equity and distribution, ignoring the heterogeneous nature of most communities. Using the case of a small region of Kullu district in the Western Himalayan state of Himachal Pradesh in India, I will look at some of the problems of the concept of indigenous knowledge. However, I will also show that the possible solutions to sustainable natural resource management in the region lie not in throwing out the concept of indigenous knowledge but in linking it unquestionably with local empowerment.

Indigenous knowledge in the Sainj valley

The Western Himalayas are one of the biodiversity hotspots of the world today. With a fast disappearing forest cover and several rare, endangered and endemic species, this is a region that is increasingly attracting the attention of conservationists. Himachal Pradesh is an Indian State that lies in this mountainous region. The State has three major agro-ecological zones influenced by altitude, a lower sub-tropical zone, a middle moist temperate belt, and higher alpine and snow-covered regions. Like most of India it is well populated with a population density of 55 persons/sq km on the average. Much of this population is concentrated in the lower and mid-altitude regions. Over 35 per cent of the State's geographical area is classified as forests. Agriculture and forestry form the backbone of the State's economy since there is very little industrial development. Tourism has emerged as a significant influence in the last decade on both the economy and the environment of this region.

Kullu is a central district in the moist temperate region of the State that is rich in both forests and people. It is a district known for the beauty of its natural resources and its colourful people and for its festivals and fairs. Manali, one of the two major tourist towns in the State, lies in this district. Within Kullu, I will focus on the communities in the Sainj valley and their use and management of various natural resources in this region. The Sainj valley branches off from the main Beas valley that is the centre of this district. It encompasses the Sainj river that eventually joins the Beas, Kullu's major river system. It is an area rich in floral and faunal diversity typical of the Western Himalayas, and parts of the Sainj region are included in the Great Himalayan National Park area.

Indigenous knowledge is a term that has been used in a wide variety of connotations.[1] In its broadest sense, it simply refers to the knowledge available in a local area. Indigenousness implies something that is location specific in its origin, in its nature or in its use. Indigenous knowledge is the knowledge of a specific group of people who live in a distinct geographical or social community knowing more about their specific ecological, geographical, social, economic and cultural environment than any outsider. In this sense, the knowledge that I describe in the Sainj valley in the Himalayas is definitely indigenous knowledge.

However, the proponents of indigenous knowledge have often sought to describe indigenous knowledge as an isolated system of knowledge in remote areas, distinct and different from Western scientific knowledge.

> Science is an open system whose adherents are always aware of the possibility of alternative perspectives to those adopted at any particular point of time. Indigenous Technical Knowledge, on the other hand, as a closed system is characterised by a lack of awareness that there may be other ways of regarding the world.
>
> — Howes and Chambers 1980.

This type of definition implies a secluded group of people unaware and unconnected with other societies. Researchers are increasingly recognising that such isolated and autonomous societies rarely, if ever, exist in reality.

The communities of the Sainj valley along with Jiwa and Tirthan valleys are the most remote in this district, and these valleys have each developed distinct traditions and customs

---

1. The literature on the meanings and usages of the term indigenous knowledge is vast. See Sillitoe 1998. A number of other terms such as indigenous technical knowledge, local knowledge, traditional knowledge, etc., have also been used interchangeably in the literature for the same concept, and I will use them similarly in this paper. While recognising the problematic nature of these terms, I will, however, use them without quotation or italics, for ease of reading.

(Harcourt 1871). The area is mountainous and several villages are not connected by paved roads. Many villagers still have to walk several hours or even days to reach an all weather road or a bus stop, and are often cut off during the monsoons and in winter. These river valleys have suffered the least impact from the tourist influx that has changed the profile of the Beas valley in the last two decades. However, it is important to note that this does not imply that there is or was no contact with the outside world.

Remoteness is a relative term. This region has had trade contact with Punjab, Lahaul, Tibet and Nepal even in the nineteenth and twentieth centuries, and the forests of the region have been managed since the last century. Today, most villages have electricity and many of them have televisions. The large shining dish antennas that dot the valley pronounce the fact that some households even have access to multi-channel television broadcasts *via* satellite connections. Local people are well aware of societies and cultures other than their own, even those on the other side of the globe.

Indigenous knowledge is an evolving and living knowledge, constantly interacting with knowledge of other peoples. Indeed, as Agrawal (1995) points out, indigenous and Western knowledge have been interacting since at least the fifteenth century (Abu-Lughod 1987-88, 1989; Wolf 1982; Eckholm 1980; Schneider 1977; Wallerstein 1974, 1979 a and b). Indigenous knowledge may have its origin and use grounded in a locality, but it interacts, adapts, transforms and appropriates all knowledge that it comes into contact with. It is difficult, if not impossible, to isolate the knowledge generated entirely within the community to that borrowed from other societies.

Natural resources in this region have been actively managed and utilised by local people as far back as historical records exist. Like all societies in mountain regions, local people here have learned over generations to adapt to the requirements of the mountainous and forested terrain. These communities have their own local floral and faunal taxonomy (Pandey and Wells

1997:1284). Evidence of such long-standing natural resource management practices in this region is evident from edicts (an order sent out by the local rulers), copper plate inscriptions, and reports on land and forest settlements (Gazetteer 1918, Harcourt 1871, Anderson 1886). Many plants and plant products in the Sainj valley have been collected and used by local people for over a century. The early colonial forest settlements of the last century mention several of these plants and they settled the rights of local people to collect these plants for their needs (Anderson 1886). The communities of Sainj represent a group of people with a history of successfully managing their own natural resources for several centuries with a wealth of knowledge that emerges from their experiences, both local and non-local. Thus, they are a good case to understand the dynamic nature of local knowledge and its uses.

Local healers and disappearing knowledge systems

Not all knowledge is common to the entire community. Specific groups of people have highly specialised knowledge that helps them to perform particular functions in society, earn their livelihood, and/or accumulate wealth. Local doctors, graziers[2] and young men involved in the medicinal plant trade are three groups that have particularly specialised knowledge on the growth and habitat characteristics of plants, and their particular beneficial or harmful properties for humans and livestock. Some of this knowledge is under threat of extinction, while others are ascendant. The current utility of the knowledge to the community has a major role in the perpetuation or abandoning of this knowledge.

Vaids are the traditional healers in the region, who are a small group with specialised knowledge about the medicinal values of local plants. *Vaids* were the only form of medical care available to local people prior to Independence, and members of

2. I use this term to refer mainly to those who manage sheep grazing, although it would apply to other livestock managers as well.

the older generation in these villages still vouch for the usefulness of this local medical system. *Vaids* use different combinations of herbs that are available in the Sainj region to treat various ailments. These prescriptions and medicinal recipes exist only in the memories of the *vaids* and are almost never written down. Much of this traditional knowledge is passed on by apprenticeship to the next generation. Local people still use traditional medicines for common but less serious ailments like mild stomach irritations, coughs, headaches, etc.

Local and modern medicines are seen as complementary and people often resort to both simultaneously. However, few people rely only on traditional medicines for serious ailments. This is local knowledge that is under imminent danger of disappearing as there are currently almost no apprentices to any of the *vaids* in the region. While there may be multiple reasons for this, the lack of a steady demand for the services of the *vaid* among local people is a major one. Most people I interviewed expressed respect for local *vaids*, and commonly-used several local medicinal plants. However, they did express more faith and a higher success rate for 'modern' medicine that was now available from the local hospital.

This brings up the much debated conundrum of whether and how disappearing indigenous knowledge can be saved. While some have argued for and attempted the recording of such specialised knowledge like that of the *vaids*, it poses several ethical and practical problems. The questions of intellectual property rights, distribution and assignation of rights and benefits, the beneficiaries of and the motives behind such an endeavour remain unanswered. Local people have indeed found the optimal solution for their own needs in using a combination of medical knowledge. While the modern medical system and inoculation of children and infants provides them with better immunity against several diseases, local medicines serve as a fall-back system that provides the immediate and inexpensive help that is not available in these villages.

Table 1: Some forest and alpine plants commonly used by local people (other than for timber, fuelwood and fodder)*

Local Name	*Local Use*	*Botanical Name*
Banajwain	Edible, and cures stomach aches	*Thymus sechyllum*
Bhojpatr	Used as paper in ancient times	*Betula utilis*
Bichubutti	Edible	*Gerardiana cadicans*
Chorah**	Used as a spice for tea; also helps digestion	*Angelica glauca*
Dhoop**	Incense	*Jurinea macrocephala*
Guchi**	Used as a vegetable prior to market development	*Morchella esculenta*
Kadu**	To cure skin ailments	*Picorhiza kurrooa*
Kakkar-singi**	To cure coughs	*Pistacia integrima*
Kuth	Cures stomach ailments	*Saussurea costus*
Patish	To treat skin ailments	*Aconitum chasmanthum* *Aconitum heterophylium*
Phool	To cure stomach aches	*Usnea Lucea*
Shathjalpadi/ Shat ladi	To cure stomach aches	*Ainsliaea aptera*
Shingli-Mingli**	Detergent	*Dioscorea deltoidea*
Tejpatta	Condiment	*Cinnamomum tamala*
Thuth**	Cure for rheumatism	*Salvia spp.*

* Data compiled from interviews conducted by author with local herb collectors and local doctors, and from research conducted by a local non-governmental organisation (Society for Advancement of Village Economy: SAVE).

** Has also commercial value. Many other marketable herbs are found and collected in this region. This table includes only those plants used by local people non-commercially.

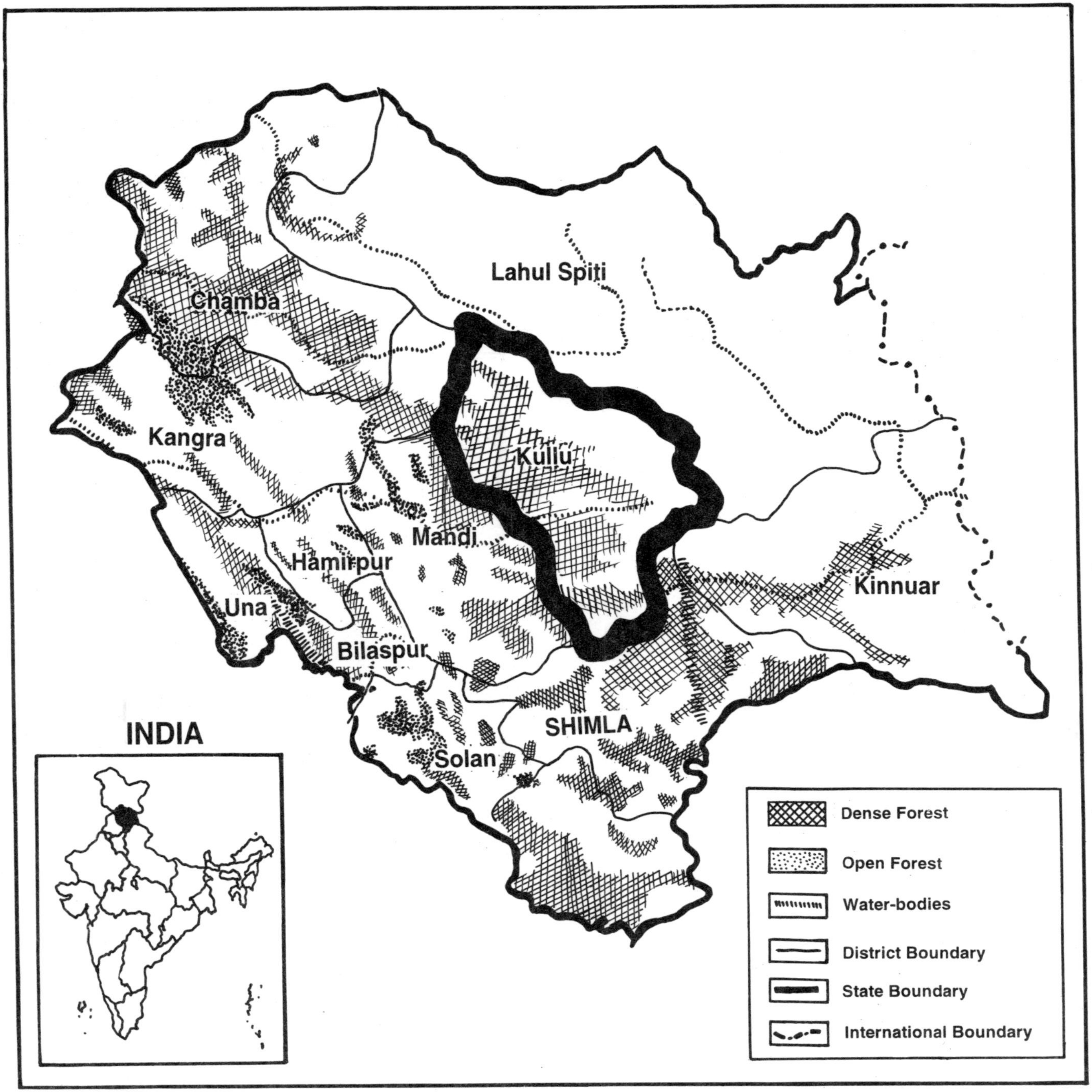

Forest Cover of Himachal Pradesh

Recording indigenous knowledge is a project that may be useful to some as a historical-cultural record. However, it is least likely to serve the purpose of assisting and/or empowering local people. Often, local knowledge is intricately integrated within social relations and institutional arrangements, and is likely to have an entirely different impact outside this context. The example of graziers discussed next is one such system in the Sainj valley where knowledge and functioning of local norms, customs, and social relations has a significant role to play in the advantages gained from local knowledge.

Holistic resource management and local knowledge

Grazing of cattle and sheep in Kullu is a traditional practice that has been much maligned as the cause of ecological devastation in this region. Nevertheless, it is an integral part of the agricultural livelihood. A group of nomadic sheep graziers (gaddis), or villagers who own large herds of sheep (fuwals) usually take most of the village livestock to high altitude alpine pastures during the summer months. For over a century, these graziers were held responsible for alleged severe degradation of the forests and pastures in this region, but recent studies have demystified ideas of graziers as environmentally destructive (Saberwal 1997). Instead, it has been shown that graziers have considerable knowledge of their ecosystem with which they interact. They are knowledgeable about the variety of grasses found along their route, and can identify plants that are particularly nutritious or detrimental to their livestock. Again local graziers have found the viable middle ground between the use of local and Western knowledge. While they use herbs to treat some livestock health problems, they also inoculate their livestock and carry Western medicines during their migration to higher altitudes.

Migratory grazing of local livestock plays an important and often neglected role in managing vegetation at lower altitudes. Moving of the village livestock to higher altitudes during the summer allows time for the regeneration of grass and understory in forests and grasslands at lower altitudes. The timing of movement of graziers closely follows the agricultural cycle. The graziers usually pass through Sainj villages after the snow melts. There is a symbiotic relationship between graziers and agriculturists. Farmers provide them with food and accommodation in return for enclosing their sheep in the fields for the night. The fields benefit from the nutritious manure at night. The graziers do not return from higher altitude meadows until after the harvest, particularly the harvest of grass after the rains, which is dried and stored for the winter.

Grasslands and forest undergrowth are systematically managed for fodder. For instance, some Panchayats (groups of villages) have fixed grazing seasons. Open access grazing is allowed on all non-sown lands from October to February. However, March to June is the season for regeneration and grazing is not allowed. People either send their livestock to high altitude pastures or stall-feed them with the previous year's stock of hay or agricultural residue. From June to October, the grass is cut from all non-sown lands and dried in preparation for the winter. This is also the time when most of the livestock is in the high altitude alpine pastures. Apart from reducing grazing pressure during these months, it also gives people the time to harvest grass and other agricultural produce without the additional burden of managing livestock. Local migratory grazing performs a role similar to rotational grazing which is popular in modern range management. It provides grazing lands at lower altitudes time to regenerate. The system, however, cannot function without the social network that exists between farmers and graziers. The route followed, the resting areas for livestock, and the timing of migration are all dependent on this mutually beneficial relationship. Such informal local institutions are difficult to identify and specify or recreate, but are crucial to the functioning of indigenous knowledge.

Sacred groves: power, equity and local knowledge

There are several formal traditional institutions that play a significant role in conservation of certain natural resources. Sacred groves are an example of such a traditional system of natural resource management. This is a concept where a definite bounded area is defined by the community as being 'sacred' and resource use in this area is severely restricted. This system has been reported in several parts of India (Gadgil 1975; Gadgil and Vartak 1976; Induchoodan 1991, WWF 1996). Sacred groves are of particular significance because conserving smaller patches of an ecotype rather than having large national parks is increasingly being seen as a more viable option by many natural resource managers in India. In Himachal Pradesh, sacred groves called Dev Vans are found extensively in Kullu and Mandi Districts.

In many areas of Himachal Pradesh, sacred groves are oases of possibly the oldest standing vegetation in the largely young forest or agricultural landscape. The size and quality of forest conserved varies widely, and the restrictions of use also vary at each site. One of the densest patches of moist temperate forests in the Sainj valley with cedar trees of the largest girth is found in Raila Panchayat. This patch of forest has been conserved by local people as the Raila *dev van*. No harvest of any kind is allowed in this forest, except wood harvested annually to repair the temple of the presiding deity of the forest. Even the collection of fuelwood and fodder or the collection of dead and fallen wood is prohibited in this forest. Local people believe in and fear the wrath of the presiding deity if these rules are broken. They belief in the *devta* and his/her power of retribution plays a major role in the conservation of the *dev van*.[3]

---

3. Many local people recount the story of a forest guard who once disbelieved and disregarded these rules. He entered the forest when he was drunk and spoke disrespectfully of the deity. As expected, he suffered great personal problems, his son fell seriously ill, and he was transferred. He finally realised his mistake and apologised to the *devta*. The exact story of this transgression and the punishment varies as different people narrate it, and no one could identify the exact forest guard or the timeframe of the alleged incident. However, the moral drawn is always about the power of the local *devta* and misfortunes that will befall those who transgress the rules of the *dev van*.

The rules regarding the use and management of sacred groves is often presented as a collective communal decision. Popular sentiment does play a part in these management decisions, but the final authority lies in the *devta* committee. The *devta* committee is a major centre of power in Kullu villages, where belief in the *devta* pervades all activities in the village. The committee usually consists of a *bhandari* or manager of the temple property including the sacred grove, a *kayath* or cashier, *a pujari* who conducts the rituals, one or more *gurs* who act as the oracles receiving and conveying messages of the Gods, a *bhandari* or storekeeper, and several *bajantris* or musicians. Except for the musicians who are lower caste men, all the other members of the committee are upper caste men. The positions on the *devta* committee are held by men of the most powerful households within the village.

Indigenous knowledge and local systems of management that utilise this knowledge have often been idealised as being people-friendly and equitable. Some authors have even argued that indigenous knowledge and management systems empower and privilege women (Shiva 1988). However, as the constitution of the *devta* committee suggests, this is often not the case. The equitableness of the rules imposed and the benefits obtained vary widely within the valley. Most remaining *dev vans* in this region are found in the higher altitude regions, where there is generally more forest cover. Many sacred groves in this region allow all uses of the *dev van* except timber harvest. In many areas of Himachal Pradesh, fuelwood is often a critical problem for lower income households (Vasan 1996). Lower income households are more dependent on forests for non-timber products. In the case of higher altitude *dev vans*, like Raila *dev van*, this is not a major consideration. There are fairly good forests adjacent to the *dev van* that fulfil the needs of fuelwood, fodder and other forest products. However, in less forested regions, the shortage of fuel and fodder is felt most acutely by lower income households. In addition, many *dev van* particularly prohibit women responsible for fuel and fodder collection from entering the forest.

The concentration of power in a biased *devta* committee is not only inequitable locally, but is also a threat to the future existence of sacred groves in this region. Several *devta* committees have decided in recent years to harvest the sacred groves and invest in land or ornaments for the *devta* or construct new temples. The committee has more flexibility and fluidity in managing property in the form of ornaments rather than as forests. In the adjacent Tirthan valley, a *devta* committee had decided during the last year to harvest a sacred grove and had even contacted several forest logging contractors. However, at the last minute other villagers intervened and prevented the committee from carrying out the harvest. While the *devta* committee wanted to harvest the forest, the rest of the village wanted to conserve it and in this case succeeded in enforcing their will.

While sacred groves are an important traditional institution, their continuance depends on the continued belief of local people in this system. It does not depend as much on the belief of outsiders in this system. In the present democratic society with an increasing awareness of rights and equity, sustainability of natural resources can be ensured only by the encouragement of more equitable management systems.

Indigenous knowledge and markets

The development of a vibrant market economy has definitely affected the management of natural resources and of indigenous knowledge in the region. As early as 1872, a colonial settlement officer writes

> (Local people in Kullu could) earn a good deal of money ... by gathering wild roots and herbs in the forests and selling them in Sultanpur or at Rampur for export to the plains.
>
> — Lyall 1872 cited in Singh 1998.

Although trade in natural products has a very long history in this region, the scale of this trade has increased tremendously since transportation and communication have improved in recent times. One of the most severely affected resources are the forest and alpine herbs that have medicinal and cosmetic value. Herb collection for the pharmaceutical industry is a typical example of how market expansion uses and overcomes traditional knowledge. Several herbs found in the Sainj region are in great demand both within and outside the country. This has become a major source of cash income in the region, and contributes up to 75 per cent of the cash income of some households.

The primary collectors are the young male members of the community. Collectors possess various information, knowledge and skills that are necessary for this occupation. Apart from identification and extraction methods, most collectors are also cognisant of the altitude and habitat conditions under which different herbs occur. They know the method of propagation and seasons of fertilisation of the particular plants that they collect. Most of this knowledge is gained through informal apprenticeship, as collectors generally travel in groups, and the first time collectors learn from experienced community members. Knowledge of the terrain and areas where the herbs are available is a distinct advantage of local collectors.

Earlier, community rules ensured that sustainability was maintained by restricting collection before the fertilisation season. For instance, experienced collectors recount that until a decade ago the collection season for alpine herbs did not begin before the first of June. Before then, most herbs are immature, and larger numbers have to be collected to complete a single load. Unripe herbs are often of lower quality and command a lower price in the market. However, with increasing market demand people start collecting as soon as the snow melts in May, giving the herbs little time for flowering or maturing. Markets dictate the number of collectors and the collection time. Chape Ram, a herb collector for the last twelve years, gives an example:

> The best market for *Dhoop*, which is used to make incense sticks, is in June because wholesalers are

> generally out of stock by this time. The previous year's stock is already sold out and there is demand for more. Best *dhoop* roots are found in September or October after the rains. However, by then, the market is flooded with *dhoop* and prices are very low. So, we don't collect it after rains anymore. Also, if no one harvests in an area for 2 or 3 years then we get very thick roots and our work is easy. But nowadays it is difficult to find a patch that has not been harvested for even one season.
> — Pers. Interview, June 1998.

Collectors are well aware of the concept of rotational management and the need for regeneration. However, as the market demand for these products grows, the social rules that utilised this knowledge are gradually eroding.

On the one hand, this is an example of the erosion of indigenous systems of resource use by growing external markets. On the other, it is an opportunity for local people to derive economic benefits from their knowledge and natural resources. Local knowledge of regions where herbs are available and mountainous routes is effectively combined with new knowledge about the market for these products. This is another aspect of the transformation of local knowledge and resource management systems.

Local knowledge and natural resource management

It is accepted wisdom in conservation and development circles today that people tend to conserve resources in which they have a significant stake. However, with indigenous knowledge, when planners and policy-makers praise its utility, it is often situated in the past rather than the present reality. Its adaptation and use in the changing circumstances of society has often been criticised and criminalised. This leads to surreptitious illegal use as well as to middlemen benefiting due to the lack of free flow of market information. Much of the criticism and rejection of indigenous knowledge can be traced back to overzealous attempts to isolate it from modern day realities. Understanding indigenous knowledge implies not only knowing and recording the technical aspects of this knowledge but also understanding the social aspects of it in a changing environment. Idealistic visions of local people being satisfied with bare subsistence levels of production have led to projects that work against any form of economic incentives for local people.

Local knowledge in the Sainj region is a combination of traditional knowledge and management systems that have managed resources over generations, and knowledge of the changes in the transport, communication and market demands. A major reason for the conservation of the natural resources of this region has been the realisation by local people of the benefits of their resources. One method of conserving natural resources as well as indigenous knowledge is to make it more useful to people's current needs.

For instance, conservationists recommend the cessation of all herb collection in the region, since the collection is ecologically unsustainable.[4] As expected, there is little support for this idea among local people, and it is very difficult to implement. A more viable option is to setup collection cooperatives for the sustainable collection of herbs. These are likely to be more sustainable ecologically and socially, and more manageable than a total ban on collection of herbs. The cooperative would function to ensure that collection is maintained at sustainable levels, as well as that local people benefit more from these resources. The methods and times of collection should be decided upon by the cooperative of local people depending both on sustainability issues as also on labour availability and family cash needs. Such a system requires that policy-makers respect and trust local people's capacity to use their knowledge and resources optimally.

---

4. Sustainability is a concept that defies an exact definition, and is often the site of contest. I will, however, use the term to imply a level of resource use that is negotiated among the various stakeholders, and ensures a fairly long-term continuation of the current state of the resource.

Environmental conservation efforts have gradually begun to shift away from law enforcement and use restrictions towards more participatory approaches emphasising local communities and their needs (McNeely and Miller 1984; Pandey and Wells 1997). Many supporters of sustainable development or conservation today accept that for a project to succeed, local people must be involved, not just in supplying labour but also in making crucial decisions. Projects must be based on people's own ideas and knowledge of their farming systems. Such an approach will change the ownership of projects from development agencies to the people (Gata 1993). These changes portend a new role for local knowledge within the context of these projects.

Indigenous knowledge can be used to identify local needs, design local institutions, and find local solutions. It is often the best and sometimes the only source for constructing the ecological and social history of a region. This can help in problem identification and in setting priorities. Local people also best understand the planning of time for various activities according to their own seasonal activities and conditions and are also often the best critiques of ideas that involve local issues.

There are two major projects currently being implemented in the Sainj area that focus on the involvement of local people in the management of forest resources. The Eco-development project funded by the World Bank functions in and around the Great Himalayan National Park region.

> The (project) plan proposes that National Park staff play a key role in involving the local communities in information-gathering and decision-making to facilitate participatory management of natural resources in the Project Area.
>
> — Pandey and Wells 1997

The Joint Forest Management Project funded by Department For International Development (DFID) encompasses Kullu and Mandi districts of Himachal Pradesh. The objective of this project is also to involve local people in the management of local forests.

Both projects have selected villages in the Sainj area, and have a mixed record of success.

These are projects in the right direction in their objective to involve local people and utilise their wealth of knowledge in the common goal of sustainable development. However, questions of power in decision-making remain. Such projects tend to have their own agendas such as "finding alternatives to forest use in the National Park area" or managing of "degraded protected forests" whereas local people may be more interested in finding better markets for forest based products or being involved in managing undegraded forests in their area. These alternatives often involve a higher degree of risk for project planners because sharing of power implies that local people will also make decisions about managing their own resources that may be unpleasant to us (more powerful people, policy-makers, funding agencies or conservationists).

More optimistically, the only forests that do exist today in India exist where local people have decided to manage and protect them. More policy-makers the world over are realising that ecological resources cannot be conserved without the cooperation of those who have lived near it and managed it for centuries. If the people who have a stake in their own environment, their own economy and their own knowledge, have the power to act on their knowledge and needs, by common reason the outcome would be one where the environment, economy and knowledge are best utilised. This is the ultimate promise of indigenous knowledge.

References

Abu-Lughod, J. (1987-88), The Shape of the World System in the Thirteenth Century, Plus Comments, *Studies in Comparative International Development*, 22: pp. 3-51.

Abu-Lughod, J. (1989), *Before European Hegemony: The World System A.D. 1250-1350*, New York: Oxford University Press.

Agrawal, Arun (1995), Dismantling the Divide Between Indigenous and Scientific Knowledge, *Development and Change*, 26: pp. 413-39.

Anderson, A. (1886), *Rights and Settlements of Kullu District*, Lahore: Civil and Military Gazette.

Brokensha, D., D.M. Warren and O. Werner (eds.) (1980), *Indigenous knowledge systems and development*, University Press of America.

Eckholm, K. (1980), On the Limitations of Civilization: The Structure and Dynamics of Global Systems, *Dialectical Anthropology*, 5: pp. 155-66.

Gadgil, M. (1975), Sacred Groves of India, *Journal of the Bombay Natural History Society*, Vol. 7.

Gadgil, M. and V.D. Vartak (1976), Sacred Groves of the Western Ghats of India, *Economic Botany*, 30: pp. 152-60.

Gata, Ntombie (1993), Traditional Knowledge and Agricultural Sustainability, In: Davis S.H. and K. Ebbe (eds.), Traditional Knowledge and Sustainable Development, Proceedings of a Conference held at The World Bank, Washington, D.C., September 27-28, 1993.

Gazetteer of Punjab (1918), Vol. XXX District Kangra, Kulu, and Sainj. Lahore: Government Press.

Harcourt (Captain, later Colonel) A.F.P. (1871), *The Himalayan Districts of Kooloo, Spiti and Lahaul*, Reprinted in 1982, New Delhi: Vivek Publishing Company.

Howes, M. and R. Chambers (1980), 'Indigenous Technical Knowledge: Analysis, Implications and Issues', in D. Brokensha, D. Warren and O. Werner (eds.), *Indigenous Knowledge Systems and Development*, pp. 329-40. Lanham, MD: University Press of America.

Howes, M. and R. Chambers (1979), Indigenous technical knowledge: analysis, implications and issues, *IDS Bulletin*, 10(2): pp. 5-11.

Induchoodan, N.C. (1991), *Sacred Groves of Kerala*, New Delhi: WWF.

Lyall, J.B. (1872), *Kangra Settlement Report* (completed in 1872, jointly published version of settlements made by G.C. Barnes and J.B. Lyall, Lahore, 1889).

McNeely, J.A. and K.R. Miller (eds.) (1984), *National Park Conservation and Development: The Role of Protected Areas in Sustaining Society*, Washington, DC: Smithsonian Institution Press.

Pandey, S. and M.P. Wells (1997), Ecodevelopment planning at India's Great Himalayan National Park for biodiversity conservation and participatory rural development, *Biodiversity and Conservation,* 6: pp. 1277-92.

Saberwal, Vasant K. (1997), Pastoral politics: bureaucratic agendas, shepherd land-use practices, and conservation policies in Himachal Pradesh, India, 1865-1994, Doctoral Dissertation, Yale University, New Haven.

Schneider, J. (1977), Was There a Pre-capitalist World System? *Peasant Studies*, 6: pp 20-27.

Shiva, V. (1988), *Staying Alive: Women, ecology and Survival in India*, New Delhi: Kali for Women.

Sillitoe, P. (1998), The Development of Indigenous Knowledge. A New Applied Anthropology, *Current Anthropology*, Vol. 39 (2): pp. 223-51.

Singh, Chetan (1998), *Natural Premises. Ecology and Peasant Life in the Western Himalaya 1800-1950*, Delhi: Oxford University Press.

Vasan, S. (1996), Gender and Class: Intersecting Variables in Forestry, *TRI News*, Yale University, Spring 1996.

Wallerstein, I. (1974), *The Modern World-system*, I, New York: Academic Press.

Wallerstein, I. (1979a), *The Modern World-system*, II, New York: Academic Press.

Wallerstein, I. (1979b), *The Capitalist World Economy*, New York: Cambridge University Press.

Warren, D. M. (1991), 'Using Indigenous Knowledge in Agricultural Development', World Bank Discussion Paper 127. Washington, DC: World Bank.

Warren, D.M., G. von Liebenstein and L. Slikkerveer (1993), Networking for Indigenous Knowledge, *Indigenous Knowledge and Development Monitor*, 1(1): pp. 2-4.

Warren, D.M., Slikkerveer, I.J. and Titilola, S.O. (eds.) (1989), Indigenous Knowledge Systems: Implications for Agriculture and International Development, *Studies in Technology and Social Change*, No. 11. Ames, Iowa: Technology and Social Change Program, Iowa State University, USA.

Wolf, E. (1982), *Europe and the People Without History*, Berkeley: University of California Press.

WWF (World Wildlife Fund) (1996), *Sacred and Protected Groves of Andhra Pradesh*, New Delhi: WWF.

10

Man's Relationship with Forest — Deification of Trees and Plants

Harish Chandra Das

Evolutionary aspects of man and forests

IT IS universally accepted that trees are among the oldest and longest living things on earth. Animals and human beings were later creations. The new born baby first saw the beauty of the forest as the primitive mother brought forth her first child under the shade of a tree. To protect this baby from animals the mother used the branches of trees as its cradle. It got a tree as a comrade. Neither of the comrades could speak, still they maintained their friendship. From the beginning the tree became an indispensable part of the primitive man's life.[1] When human beings emerged in the process of evolution they found themselves in the lap of nature. The primordial forest consisting of trees, plants, hills, dales, streams and rivers was the life-force of the animal world. The practical benefits of the world of vegetation have been enormous. Plants and trees have been the source of shelter, comfort and food to animals and human beings. The multi-medicinal properties and life-giving force of the plants and trees, the variegated leaves, flowers and fruits, the woodland with green foliage vibrating with the melodious notes of birds perched

1. Sunkur Sen Gupta (ed.) (1965): *Tree symbol worship in India. A new survey of a pattern of folk-religion*, Indian Publication, Calcutta. p. xxii.

on branches, the roaring sound of the animals in the mountain dales, the jingling sound of the flowing streams, much more important and charismatic, its life-giving force as the mother led the people to develop the feeling of adoration and deification of the world of vegetation. This process has continued unabated to the modern age.

In the first phase of reflection, primitive man leading his wandering life in the forest, mountain and river valleys, observed with awe and wonder the sun, moon, sky, stars, rain, thunder, lightening, sea, river and other natural phenomena attributing to them supernatural powers. Over the passage of time adoration for natural phenomena was universally accepted. In regard to the Hindu religious view of trees and plants the views of Dr. T.C. Majupuria are referred to here:

> According to the Hindus, God, the Supreme of the supermost, the Mighty of the Mightiest, is the giver, and human beings, the inferior creatures, are the recipients of the gifts of God. So any being or object whose action is to impart, to give, to offer benevolently, charitably, unhesitatingly with no motive for anything in return, is reckoned as God in our common parlance. Thus our trees testify that they, right from the primordial period of the universe, have been extending shelter to the passers-by, exhausted and dissipated, light and knowledge to the sages and noble-hearted persons who practise austerity and penance, the roots and fruits to the people and animals in wild surroundings, and shelter to the birds which build their nests in the foliage. The trees, as they have been the givers, are therefore, naturally the representatives of God, if not the God in themselves, and this must have been the reason that the Oriental saints, sages and philosophers came forward to offer the highest salutation and sincerest respect to the

trees with a message to the coming generation to follow in their foot-prints.[2]

History delineates that at the dawn of reflection, there were immense primeval forests of various types. All ancient cultures beginning from the pre-historic to historic age flourished in the midst of the forests and on the banks of rivers. The primitives whom we term 'Adivasi' or 'autochthonous' developed their cultures amidst dense forests in a naturalist manner uncontaminated by the sophisticated societies. Born and brought up in the natural surroundings they developed the highest sense of respect and veneration for the forest. Saints and seers spent their life in the forest hermitage for religious and philosophical pursuits. The knowledge they bequeathed through the *guru-shishya-parampara* enriched Indian philosophy and religion.

PRE-VEDIC PERIOD

In India man's relations with the forest and tree worship are traced from the Indus valley civilisation. Several seals from Mohenjo-Daro and Harappa depict tree gods and goddesses with supplicating figures in front of them. In one such seal a Pipal tree (*Ficus religiosa*) with two branches is represented, and in between the indwelling appears in nude form with triple horns and armlets. There are seven standing miniature figures with a long plant of hair rolling down the neck and a plume on the head. They appear to be the ministrants of the goddess. In this way there are various representations of sacred trees carefully carved by the artists. This clearly establishes the fact that the worship of trees in the indigenous way was prevalent in India before the advent of the Aryans.

VEDIC PERIOD

During the Vedic period agriculture, medicine and arbori-

2. T.C. Majupuria (1988): *Religious and Useful Plants of Nepal and India*, Department of Medicinal Plants, Botanical Survey and Herbarium, Kathmandu, p. 10.

horticulture developed to a great extent. There is evidence to show that a large number of terms were used to describe plants and their parts, manuring and rotation of crops for improvement of soil fertility and plant nourishment. On the basis of their forms, the *Rigveda* classifies plants into *vriksa* and *druma* (tree), *vishakha* (shrub with spreading branches), *sasa* (herb), *amsumalin* (spreading plant), *stambini* (bushy plant), *vratati* (climber), *pratanvati* (creeper) and *alasala* (creeper spreading on the ground). The *Rigveda* and *Atharvaveda* give a list of different parts of the plant. A hymn in the *Atharvaveda* relates:

> Rich in sweets the roots, rich in sweets the tips of them, rich in sweets the middle of the plants (stem); rich in sweets the leaves, rich in sweets the flowers of them; rich in flowers, rich in shoots, rich in fruits, also those lacking fruits ...[3]

The *Brhadaranyaka Upanishad*[4] compares a tree to the human body and speaks of its leaves as hair; the outer bark as the skin; the sap as blood; the inner bark as the flesh; the innermost layer of bark as the nerves; its wood as the bones; and its pith as the marrow. According to the Vedic thinkers plants had preceded animals and particularly man in the process of evolution. In a passage of the *Taittiriya Upanishad* the idea of evolution is suggested:

> From that very *Atman* ether came to be; from ether air, from air fire, from fire water, from water the earth, from the earth herbs, from herbs food and from food the person came into being.[5]

Similarly the concept of soul is the foundation of religion

---

3. R.C. Majumdar (ed.) (1980): *History and Culture of the Indian People*, Bombay, Vol.II. p. 580.

4. R.C. Majumdar (1993, Reprint): *Botany in Ancient and Medieval India*.

5. R.C. Majumdar (1993): *Cultural Heritage of India*, Calcutta, Vol. IV, p. 118.

throughout the primitive society. The soul is commonly believed to have an ethereal existence, the size and shape of which differs from tribe to tribe. According to the Lakhers of Assam the soul resembles the body in appearance and size. The Kukis of Manipur conceive the soul as a replica of the individual. The Naga have the same view. The tribes of Chhota Nagpur and northern and central Orissa conceive the soul as the *chhaya* or shade of the individual. The Saoras believe that the soul (spirit) of the dead ancestors lives in the underworld and is finally freed after performance of the *Karja*[6] ceremony. The Toda of the Nilgiri hills believe that the soul resembles the individual and in the other world leads a life similar to this living one. Thus the concept of soul is a universal phenomenon in the tribal world.

POST-VEDIC PERIOD

The post-Vedic period witnessed the development in science of medicine, agriculture, arbori-horticulture and sylviculture. *Vriksayurveda* (the knowledge of tree life) *Bhesaja-vidya* (knowledge of medicine) dealt with plants and trees in a more systematic way. The *Arthashastra* of Kautilya (fourth century BC), *Agni Purana* (fourth to fifth century AD) and *Brihatsamhita* also referred to trees and plants. The post-Vedic Indians knew that air, warmth and water are necessary for germination. The term *uttanapada* used in connection with germination is important as the *pada* or root that comes out first is the process of germination. Shukracharya in his *Shukraniti* gives detailed description of various parts of plants and their importance. In the *Shukraniti* the king is compared to the root of a tree, the ministers to stems or trunks, the commanders to branches, the troops to leaves and flowers, the subjects to fruits and the land to the seed. The *Mahabharata* refers to trunks, branches, nodes, leaves, flowers, fruits and seeds of different kinds of trees. Similarly, the *Vishnu Purana* classifies the principal parts of plant into *mula* or *pada* (the subterranean part), and *tula* or *vistara* (subareal part). *Tula* or *vistara* comprises two parts, viz., *kanda*

6. Final funeral ceremony of the dead ancestors.

(stem or axis) and *parna* (leaf). Plants may be *sakanda* (with stem) and *aprakanda* or *stambha* (without stem). A bushy plant is described as *ksupa*. *Shakha*, *prashakha* and *anushakha* are the terms used for branches in descending order. *Parna* or *patra* is *savrinta* (petiolate) or *avrintaka* (sessile). A leaf may be *ekapatra* (unifoliate), *dvipatra*, *tripatra*, or *sapta-patra* and so on according to the number of leaflets. The shape of the leaf is also termed as *ashvaparnaka* (like the ear of a horse), *musikaparni* (resembling the ear of a mouse), *kisaparna* (like a monkey's ear) and *hamsapadi* (like the foot of a duck), etc. The flowers and fruits have also been classified. The post-Vedic literature refers to different kinds of plants. The creeper is termed as *lata*, *valli* or *vratati*. The five regions in the body of the plant are termed as *tvac* (skin), *mamsa* (soft tissues or bast), *asthi* (wood), *majja* (pith) and *snayu* (fibre in the bast). The *Shukraniti* and Sankara Misra's *Upaskara* refer to healing up of wounds by natural recuperation.[7] In this way the ancient texts deal with arts of seed collection and selection, soil selection, sowing, seed germination, propagation such as grafting and cutting, planting, nursing, manuring, crop rotation, plant treatment, plant classification and identification, landscaping, etc. In connection with test of medicinal quality of plants a fact is cited here. Bhiksu Atreya, the celebrated teacher of medicine in the university of Taxila, wanted to test the knowledge of his illustrious pupil Jivaka (who later on became the physician of Bimbisara) on the medicinal quality of the herbs. Atreya asked his pupil Jivaka to seek a *yojana*[8] around Taxila[9] such plants which have no medicinal value. Jivaka went out in his mission and tested whatever plants he came across, but could not find a single plant which had no medicinal value. When he reported this, the teacher was highly satisfied and declared him successful in the discipline of medicine.

---

7. R.C. Majumdar (1993): *Botany in Ancient and Medieval India*, p. 120.

8. = 24 miles radius.

9. A famous Buddhist place, now to be found in Pakistan.

Charaka and Sushruta, two experts of medieval India in the science of traditional medicine, classified lands according to the nature of the soil, climate and vegetation into three categories, viz., Jangala, Anupa and Sadharana. Jangala is described as a region of open spaces pervaded by expansive mirages having rivers, rivulets and wells. It comprises mainly dry and rough sands, khadira (*Acacia catechu*), asana (*Terminalia tomentosa*), babari (*Zizyphus jujuba*), etc. Anupa is a marshy land bordered by sea. Due to the network of rivers and accumulated water the area becomes marshy. Some of the important plants in this region are Vanjula (*Calamus species*), Hentala (a type of palm), Narikela (*Cocos nucifera*), etc. The Sadharana region has some characteristics common to the other two regions. A few of the plants found in this region are Mandara or Parijataka (*Magnolia grandiflora*), Santana (*Ficus benghalensis*), etc.

The plants are named according to their special association, medicinal and other values, morphological characteristics, environmental association and other peculiarities. Names of certain plants are associated with historical events. As for example, the Bodhidruma (*Ficus religiosa*, tree of enlightenment) received its name on account of its being a tree under which Gautama got enlightenment and the Neem tree (*Azadirachta indica*) is held sacred as the images of Jagannath, Balabhadra and Subhadra are carved from it. On account of its sacredness and association with the cult of Jagannath people of Orissa do not utilise Neem wood for any other purpose. The Suddha Saora hold Neem tree as their god Kittung (Jagannath). Morphological features were embodied in *tripatra* (three-leafed), *panchangula* (five-fingered), *satamuli* (hundred-rooted), *magadhi* (native plant of Magadha) and *campeya* (native of Campa). Charaka also classifies plants on the basis of their medicinal value and prescribes their utility in treatment of different diseases.

MEDIEVAL PERIOD

The classification of plants was more systematically done in medieval India. Of the medical treatises composed during this period Sharngadhara's *Sharngadharapaddhati* (AD 1300) is the most outstanding. *Upavana-Vinoda*, one of the chapters of the work, deals in detail with the classification of plants, selection of seeds, sowing, planting, watering, protective and curative measures, proper nourishment, utilisation of fertilisers, methods of propagation, etc. Studies of plant life with reference to environment were intensively made. The cultivation of cotton plants was emphasised. Thus study in plant life made a phenomenal progress.

The post-Vedic period saw the emergence of two important religions — Buddhism and Jainism. Buddha and Mahavira, the protagonists of these popular religions, got enlightenment in the forests under sacred trees. The Bodhi tree (*Ficus religiosa*) under which Buddha got his enlightenment came to be regarded as a sacred tree and is hence widely worshipped. Similarly the Keval tree (Banyan or Peepal) was associated with Jainism. These are the historic events which are responsible for enhancing the sacredness of some trees.

During the rule of the emperor Ashoka the sectarian religion Buddhism became the State religion spreading far and wide. The places connected with Buddha's visits were sanctified and honeycombed with Buddhist monuments. The monolithic Buddhist pillars erected in the places of interest were depicted with animals and the figures of tree worship.

In the post-Ashokan period Sanchi, Bahrut and Sarnath rose to eminence as great centres of Buddhism. The *stupa*s, *chaitya*s[10] and the railing pillars are the notable archaeological evidences depicting forest scenes, sacred horses, garden scenes, lion capitals, elephant capitals, the Buddha and dwarf capital, the *chakra* pillar, *naga* and tree capitals and scenes of tree worship.[11] These are perhaps the earliest sculptural evidences of tree and serpent worship in the first two centuries of the Christian era.

The Amaravati tope in Andhra Pradesh was another great

---

10. *Stupa*s and *chaityas* are Buddhist relic shrines.

11. J. Fergusson, *op. cit.*, plates XXIV-XXXVI.

Buddhist establishment of the sixth century AD.[12] This vast Buddhist establishment after excavation brought to light series of sculpture panels depicting the railing pillars, the decorative facets, rails of outer and closer surfaces, worship of sacred objects, life sketches of the great master, etc.

The Gupta period known in history as the golden age experienced renaissance in all branches of learning and culture. The eighteen *Mahapuranas* (religious texts), *Shilpashastras* (texts on architecture) and texts of iconography were compiled, and conceived gods and goddesses were reflected in sculptural art and temples built for the abode of such deities. The temple building activities spread throughout India under royal patronage and popular support. The temples in the post-Gupta and medieval periods were not merely the abodes of gods and goddesses but were the eminent centres of culture and religion. In the sculptural arrangement on the exterior walls the human, celestial, vegetation and animal worlds were depicted in a nicely balanced manner. On close scrutiny one can find the illustration of divinities, some divine figures, human beings in socio-religious situations, mythological scenes along with trees, plants, animals, etc. Certain trees, plants, flowers were closely associated with gods and goddesses as emblems and animals as vehicles. Goddess Lakshmi became the deity of wealth, vegetation and prosperity. Lotus is her amble; she bore the epithets of Padma (Lotus, *Nelumbium* sps.). Padmalaya (abode of Lotus), Padmaja (born from Lotus), Padmini, etc. Krishna is associated with the Kadamba tree (*Anthocephalus cadamba*), *gopis* and cattle. The Tulsi (*Ocimum sanctum*), Bel (*Aegle marmelos*), Banyan (*Ficus benghalensis*), Peepal (*Ficus religiosa*), Siali (*Bauhinia vahlii*) and Neem (*Azadirachta indica*) trees were specifically associated with gods and goddesses. The lotus flower was given the highest sacredness and was accepted not only as an emblem of gods and goddesses of the Buddhist, Jaina and Brahmanical pantheons, but it also decorated the pedestals on which the deities were installed. Such pedestals came to be known as lotus pedestals.

12. J. Fergusson, *op. cit.*, p.155.

Besides, trees, plants and creepers formed decorative motifs in the temple illustrations.

Forest and tribals

Folk religion is basically nature worship. The totemism of the tribals virtually developed from animism. This, historically speaking, was the religion of primitive people. The primitive form of religion was later on greatly influenced by the sophisticated Brahmanical religion. After the advent of the Aryans to India their religion prevailed side by side with the primitive folk beliefs. These two trends formed the arena of Indian religion, though the interchange, assimilation of the traits of both complexes were the processes of history over the ages.

It has been stated earlier that tree and serpent worship is the oldest religious phenomenon round the globe. James Frazer[13] in his famous book *The Golden Bough* has cited the examples of tree worship in India, Europe, Australia, Africa, China, Japan, New Zealand, America and other countries showing different shades of beliefs and diverse rituals and customs linked with various species of trees. In many countries sacred groves where every tree is regarded as divine are specifically earmarked in particular areas. The cutting of any tree from the sacred grove is a taboo. The spirits and deities believed to be residing in such trees are objects of veneration and hence making offerings to them is a general practice. In the primitive world everything including tree and plant is considered animate. Primitive man believes that they have souls like any living beings and felling a tree is similar to killing an animal. The concept of trees and plants as animated beings resulted in treating them as male and female who could be married to each other.

According to another opinion the tree is the abode of a spirit and it can quit and return to it at his will. Once the tree is viewed as an abode of the spirit it changes its shape and assumes the form of human being. The spirit is believed to have rightful

13. J.G. Frazer (1950): *The Golden Bough*, (Abridged edition), London, pp. 109-15.

possession over the tree finally becoming a forest god. Frazer illustrates that trees considered as animate beings are credited with the power of making the rain fall, the sun shine, the crops grow, flocks and herds multiply and blessing women with offsprings. Exactly the same powers are attributed to tree gods conceived as anthropomorphic beings or as actually incarnate in living man.[14]

The tree cult is also associated with the practice of totemism by a large number of tribes. Totemism is generally understood as a mystical relationship between a human group and some class of natural phenomena, usually a plant or an animal species. Frequently in a primitive society, members of a particular lineage or clan consider themselves descendants of the totem and treat the totemic plant and animals ritually. Goldenweiser opines, totem and totemites may be supposed to share physical and psychological traits and the totem may be looked upon as a sort of guardian angel of the totemites.[15] According to Tylor, "The primitives believe that after a person's death, his soul takes abode in certain plant or animal species and so the entire species is protected." Thus, totemism, according to Tylor's explanation emerges "... as a form of religious worship, to be more precise ancestral worship..."[16] So far as tribal India is concerned, most of the ethnic groups claim to have mystic links with the totemic objects — plants, animals and other things.

The Santal clans are named after plants, animals and mysterious objects. The Hos have more than fifty exogamous totemistic clans. Similarly, the clans of the Oraon, Kharia and Munda have exogamous clans. Out of the total series the Oraon have eighteen vegetable totems named after trees like Bar, Basa, Kindo Putri, etc. They believe that the totemic plants or animals

---

14. K. N. Sahay (1965): *Tree cult in tribal culture. Tree symbol worship in India*, Calcutta, p. 61.

15. Goldenweiser, A.A. (1910): Totemism. An Analytical Study, *Journal of American Folklore*, XXIII, No.88.

16. D.N. Majumdar and T.N. Madan (1961): *An Introduction to Social Anthropology*, Bombay, pp. 122-3.

help and protect the clan members in many ways. The people show reverence to them and do not destroy them under any circumstances. Some of the clans of the Santal are named after some plants such as Bel tree, Banyan tree, Pipal tree, etc. The destruction of such plants is taboo for that particular totemic group. Some of the clans of the Bhil are named after Bamboo (Vascula), Mango tree (Markam), Teak tree (Tekam) and Mahua tree (Irpachi).[17]

Nature of Tree Cult Among the Tribals

Tree cults among the tribals can be understood from their myths and legends, their association with the forest for subsistence, socio-cultural ties and their rituals and ceremonies. Their close link with the forest is reflected in all facets of their culture.

SACRED GROVES

The concept of sacred grove is a world-wide phenomenon. We come across from the ancient literature to the prevalence of sacred groves in several countries. In the islands of Philippines and Malaysia certain trees are cultivated for their delicious fruits. The tribals perform peculiar ceremonies to stimulate their growth. The shaman takes a hatchet and delivers several blows on the trunk of the tree. It comes through the mouth of the shaman that it would bear fruits requesting him not to hurt it. In Japan similar customs prevail. Two male members come to the tree; one bears an axe and the other climbs the tree. Both of them are supposed to be in a trance. The man with the axe threatens that he would cut down the tree. The other in the branches of the tree replies on behalf of the tree that it would bear abundantly. Parallels are to be found in Europe.

The Dravidian and Mundari-speaking tribes of Orissa and Bihar like the Oraon, the Munda, the Hos and the Santal have each a sacred grove at the outskirts of their settlements consisting mainly of Sal (*Shorea robusta*) trees. The sacred grove is believed

---

17. Kurup, A.M. (1962): Glimpses of Life and Culture of the Tribes of Madhya Pradesh, *Vanyajati*, Vol.X, No.4, p.138.

to be the abode of spirits and deities. Offerings and sacrifices are given to such spirits for the general well-being of the village. The Oraon call the sacred grove Sarna or Jaher, and believe it to be inhabited by Chalapachecho (Old lady of the grove). She occupies the highest position among the village deities and spirits. A block of stone on the foot of the main Sarna tree represents the deity. The spring festival known as Sarhul or Khaddi associated with the sacred grove is observed in the month of Chaitra (April), when the trees are full of new leaves. On this occasion fowls are sacrificed to Chalapachecho and other deities by the village priests (Pahan or Biaga). The main objective behind this festival is to keep the villagers, cattle, crop fields and other things free from malevolent spirits. This occasion is marked for dance and music by men and women.[18] Cutting down any tree from the Sarna grove is a taboo. A similar custom is marked among the Santal during their Baha Festival.

> The Sauria Paharias and Malers of the Santal pargana who appear to be the nearest congeners of the Oraon, recognise this spirit as one of their principal deities under the name of Chal to whom sacrifices are offered under a sal tree just outside Basti, a settlement among the Oraon.[19]

The Munda have a Sarna grove in the vicinity of the village. This is believed to be the primeval forest of the tribe. Their tutelary deity Desauli and his wife Jhar Era or Mabura are believed to reside in this grove. Every village has its own Desauli whose authority is confined to the village boundary. If a man of one village cultivates land in another village he is to give offerings to the Desauli of both the villages. The worship and sacrifice to such deity is organised by the priest or shaman. These deities are also propitiated in times of calamities.[20]

---

18. S.C. Roy (1915): *The Oraon of Chotanagpur*, Calcutta, p.108.
19. K.N. Sahay, *op. cit.*, p. 54.
20. E.T. Dalton (1872): *Descriptive Ethnology of Bengal*, Calcutta, p.188.

The Hill Kharia's deity of the sacred grove (male god named Boramsal or goddess named Thakurani Sal) known as Baram resides in the grove. One or more stones installed under the Sal tree represent the deity. Baram is worshipped by the dehuri, the village priests. It is given special offerings during the hunting festival (Pardhi) in the month of Phalgun (February). All members participating in the hunting expedition pay respect to Baram before proceeding to the forest. On the occasion of Fagu Nuakhia (in Phalgun) the village priest offers Mango blossoms and sacrifices a goat or a sheep to Baram. The priest prays to the deity for prosperity and immunity from diseases and calamities. This deity is also propitiated during other festivals and ceremonies.[21]

Janka, Jhanka or Sarna are known as the sacred groves of the Dudh and Dhelki Kharia. The god of the sacred grove known as Khunt or Baghia is represented by pieces of stones placed under a Sal or Mahua (*Bassia latifolia*) tree. Jankar is connected with important festivals of the Kharias like Janka or Puja (held in the month of Phalgun), Kadleta Festival (September-October), etc. Kalo, the priest, offers flowers and fruits and sacrifices fowls in the names of the grove spirit, praying and asking him to protect the cattle, goats, human beings in the jungle from the wild beasts. On the occasion of Kadleta festival twelve fowls are sacrificed to the grove spirit. This festival is organised when the upland paddy is harvested and the wet paddy lands look green with plants.

Jahira is the sacred grove of the Hos. Jahirabura is the female spirit residing in the grove. She is believed to be the consort of Desauli and is worshipped along with her consort at the time of sowing seeds and reaping of the crops.[22]

The Santal's sacred grove (locally known as Jaher) is usually situated to the west of the village. The grove comprises a number

---

21. K.N. Sahay, *op. cit.*, p. 65.
22. D.N. Majumdar (1950): *The Affairs of a Tribe*, Lucknow, pp. 45, 212.

of tall, slim Sal trees (*Shorea robusta*) and other tree species. Two of the tallest trees standing side by side are believed to be the abodes of Maran Buru (the greatest god of the Santal) and of his female consort Jaher Buri. Other trees in the grove are believed to be inhabited by Gasae Era (female spirit) and Pargana Banga (boundary spirit) and other spirits. The selection of a new site of a settlement is determined by the spirits. These grove-spirits are propitiated by the village priests during Baha, the flower festival held in October-November, Eroksim, the first agricultural festival held in June, Iri Gundhi Nawani, the millet festival held in August and Janthor held in November. Women are kept away from participating in the festivals held in the sacred grove. No woman can climb any of the trees in the sacred grove nor may she be present when the sacrifices are made nor can she eat the flesh of an animal offered to the grove spirits.[23] Besides, many other tribes have sacred groves having deities propitiated on ceremonial occasions. It is not possible to touch upon all of them in this paper.

SACRED TREES

Besides the sacred groves individual trees are held in high esteem in the tribal societies. Such a sacred tree is also believed to be the abode of malevolent or benevolent spirits. Periodical offerings are given to them; sometimes some trees possessing the evil spirits are avoided. Every Oraon village has a Mychitkha (*Ficus religiosa*) tree locally known as Hagripipa or Barandapipa. If the rain does not come in time the women of the village take water in earthen jars and pour it on the roots of such a tree. Sometimes the priest also gives offerings for the coming of the rain. People believe that after the ceremonial offering of the water to the trees the rain is sure to come. Mahadania is a malevolent village spirit; of the Oraon. Previously they used to offer human sacrifices to the spirit, now it is stopped. A fowl or goat is sacrificed to such a spirit nowadays.

---

23. See Mukherjea (1962): *The Santals*, Calcutta, pp. 233-42.

The Desauli, the benevolent god of the Hos, is supposed to reside under a tree represented by a flat stone. This tree is taboo to the tribe. No one ventures even to pluck the leaves from the tree. This deity protects the people from diseases, calamities and epidemics and from evil eyes of other spirits. Desauli is the god of rain and harvest. In the Maghi and Hero festivals and in case of epidemics the *dehuri* (village priest) offers a hen, a cock, a goat or a buffalo as it is revealed to him in a dream.[24] The Reksha-duba, the protector spirit of the Kharjas, resides on a large Semar tree. He is believed to move with an axe in hand around the village protecting it from evil spirits. L.P. Vidyarthi refers to Lurigossaiyan and Malagossaiyan (village deities of the Malar). They are believed to live on a Pushre (*Schleichera trijuga*) tree. The deities are worshipped on the Bandana festival along with sacrifice of pig, goat and buffalo.

According to the Saora some of their deities dwell on trees. Kittung is one of these gods who is a lover of trees. He weeps when the Saora cut down trees. The other gods connected with the trees are Lobosum (god of forest cleaning), Karudsum (a god born in black-palm tree), Badansum, etc. The Saora believe some of their gods and spirits to be resting in or under trees. They consider it a taboo to cut some species of trees such as Mango, Mahua, Tamarind and some other fruit-bearing trees. The trifoliato leaves of the Bel tree (*Aegle marmelos*) are needed in divination. A branch of the tree is placed at the central pillar of the house in the funeral ceremony. The thorns are used in the name-giving ceremony of the child and a branch is placed above the door to ward off the evil spirits. It is a taboo to cut the Mahua (*Bassia latifolia*) tree as it provides food and oil. The cotton plant is believed to be the abode of Buttamboi, the female ghost who approaches young men in dreams and robs them of their virility. In the epidemic of cholera the wood of the bush is used to make a small chariot in which Maridisum is escorted from the village. The Banyan tree (*Ficus bengalensis*) is held

24. D.N. Majumdar, *op. cit.*, p. 257.

very sacred. The Saora make offerings to the Banyan at the time of the harvest festival.[25] In this way several trees are associated with religious occasions.

Bura Deo, the principal deity of the Baiga, is believed to live in the Sal tree. Goats, fowls, wine and fruits are offered to him.[26] The Kadur of Cochin believe in Bhuta, Pisacha, Gandharva and Yakshini, who harm the children. These spirits are considered evil and hence are warded off.[27] The Andaman Islanders believe in a number of jungle and tree spirits with whom are associated magical powers.

FESTIVALS CONNECTED WITH TREES

The Karama festival connected with the Karama tree (Sal, *Shorea robusta*) is associated with the god Krishna. The festival is generally organised in July-August; it is virtually a festival of women and particularly of maidens. On the day of festival the Oraon maidens go out in a group, dancing, singing and playing music to cut down and bring home the branches of the Karama tree for the festival. The village priest accompanies them with the materials of worship, anoints the Karama branches with vermilion and oil and gives offerings. The three branches of the tree are brought by the maidens in baskets accompanied by musicians, dancers and a shaman. The branches of the Karama tree are then planted in the middle of the *akhara* (dancing ground) in a ceremonious way. The men and women of the village go on dancing and singing the whole night around the Karama branches. Next morning the Karama Raja is decked with wreaths of flowers, armlets, bracelets and necklaces of wild grass and the story of Karama Raja is narrated by the priest. After that all present there make offerings of flowers to the Karama Raja. The Oraon believe that the observance of the festival secures the general

---

25. Verrier Elwin (1955): *The Religion of an Indian Tribe*, London, p. 81.
26. A.M. Kurup, *op. cit.*, p. 137.
27. U.R. Ernfield (1952): *Kadar of Cochin*, Madras, pp. 156-7.

well-being of the villagers and particularly of their brothers. Then they proceed in a procession with the Karama branches. First they go to Pahan's House and then to Mahato's and Purjar's. The women of their houses give offerings to the Karama branches including fruits and bread. Finally the branches are immersed in some stream or pond.[28]

The Karama festival is sometimes celebrated by the Santals at the time of appointing a new headman of the village and sometimes in performance of death rituals of important men. This is organised for enhancing good fortune or for warding off evil spirits. Similar to Karama, Jitia festival observed after 12 days of the Karama festival is very important. This is observed by individual families, but most of the villagers participate in it. It is usually observed by the women for the well-being of their children. The ladies on this occasion observe fasting and take a ritual bath, place cucumber leaves at the door of every room in the house, pieces of bread together with flowers and fruits in the evening of the festival. Two or three Jitia Pipal branches are planted on the courtyard in front of the house. A Bel fruit is placed over the branches. The priest in the evening recites the Jitia legend. Next morning the Jitia Pipal branches along with the Bel fruit are carried and thrown on the road by the women.[29]

The trees are also associated with the life-cycle rituals. Among the Oraon certain magical means are adopted when a women gets labour pain. S.C. Roy has recorded an interesting case study of the Oraon in this connection:

> If there is in the neighbourhood a Tamarind tree (*Tamarindus indica*) which was ever singed with lightning, a man goes to such a tree, stands against it and strips off a portion of its bark where it touches his waist. The man now goes with this bark to the door of the lying-in

---

28. S.C. Roy (1938): *Oraon Religion and Customs*, Calcutta, pp. 240-6.

29. S.C. Roy, *Oraon Religion and Customs*, pp. 248-9.

> room which is forthwith closed against him, thrusts one end of the bark through a hole in the door and remains standing there holding the other end of the bark with his hands. The women has to fasten her gaze on this bark to facilitate delivery. As soon as delivery takes place, the man is informed about it and he takes out the bark, for, otherwise, it is believed, inversion of the uterus is sure to occur.[30]

S.C. Roy also gives an account of an interesting ceremony of the Oraon through which they get procreative vigour from the Sal tree.

> Twice in a year, once about a week before the Phagu festival in March and again about a week before the Sarhul festival in April or May the boys of youth dormitories are led forth by the Dhumkuria Mahto to a secluded place away from the village. The Mahto carries a new thin Sal stick and also a Sal sapling in which a slit has been made to resemble the female organ. The boys besmear one another all over the body with red ferruginous earth. The Mahto in the meantime plants the sapling firmly on the ground. The boys of the two upper grade now spit into the sapling. The boys of the lowest grade have now to strip themselves absolutely naked and each boy in turn has to insert his organ into the saliva-field slit. All the time, the boy Mahto stands behind the sal sapling with the stick in his hand and as each boy is about to step back after performing the magic operation the Mahto strikes the boy on his thigh with the sal stick. Thereafter they go back to the village Akhra.[31]

Among some Mundari tribes marrying trees is a rite in the course of their marriage ceremony. Among the Munda the bride and

---

30. S.C. Roy, *Oraon Religion and Customs*.
31. S.C. Roy (1915), *op. cit.*, pp. 241-3.

bridegroom are taken and wedded to two trees, the bride to a Mahua tree and the groom to a Mango tree. They have to touch these trees with vermilion and they clasp and also are tied to them.[32] The Malpaharia tribe of the Santal Parganas practises a similar wedding ceremony. Accompanied by the musician and the village priest *dehuri* the groom is taken to the mango tree. The *dehuri* holds a sword in his hand and moves round the tree five times with a raw thread in his hand while the groom holds the other end of the thread to the tree. After this the *dehuri* cuts a small piece of the tree skin and gives it to the groom to taste it. A similar ceremony is also observed at the bride's house. She is married to a Mahua tree.[33]

Certain trees such as Bhelwa (*Semicarpus anacardium*), Kendu (*Diospyros melanoxylon*) and Sindhar (*Vitex negundox*) are believed to have a magical potency and hence their saplings and branches are used by the tribals to avert evil spirits from the settlement as well as from the crop field. Similarly certain medicinal plants are also associated with magical practices. Before they are used for particular purposes, rituals are observed on specific occasions.

Deified Plants

The deification of plants is a universal phenomenon. It has been indicated earlier as a religious aspect of the vegetation world over the ages. Since the history of tree-worship has been delineated in the previous sections, I deem it wise to touch upon some popular plants and trees individually to indicate their significance in the religious life of people.

TULASI OR VRINDA OR HOLY BASILICUM (*OCIMUM SANCTUM*)

The Tulasi plant commands respect and adoration all over the world. The people of ancient Italy and Greece paid reverence to

---

32. E.T. Dalton, *op. cit.*, p. 194.

33. K.N. Sahay (1956): Malpaharia Village, (Unpublished M.A. Thesis, Bihar University).

this plant as it grew on the grave of Christ. The Tulasi plant has been worshipped in the Hindu world since times immemorial. It is found in front of every Hindu settlement and Vishnu temple. It is the symbol of Vrindabati, the wife of the demon Jalandhara. Owing to her chastity and purity Jalandhara sustained his life. He tortured the three worlds. He could not be killed because of his wife's chastity. Krishna planned to destroy the chastity of Vrindabati. He appeared before her in the disguise of her husband and enjoyed her. When Vrindabati came to know the fact she cursed Krishna. As a result of his mischief Jalandhara was killed. Krishna pacified her by saying that she would be born as the sacred Tulasi plant and her leaves would be used on his head. Since then Tulasi is a sacred plant. According to the *Brahmavaibarta Purana*, Tulasi was a Gopika, one of the beloved consorts of Krishna. For her close association with Krishna, Radhika cursed her to be born as a Tulasi plant. Later on she got a boon from Brahma that she would be the wife of a demon Shankhachuda, and Vishnu out of sympathy would turn himself to a Salagram. The Tulasi plant is regarded as the incarnation of Lakshmi. Its marriage is performed everywhere with Salagram (Vishnu) and its leaves are placed on the head of the latter.

In the *Vishnuyamal Tantra* it is mentioned that the Tulasi should be nurtured for three months and should then be worshipped. On the 11th day of the bright half of the month of Karttik, the marriage of Tulasi is performed. The symbol of Tulasi should be in silver and that of Vishnu should be in gold. Tulasi should be decorated and offered to Vishnu in a marriage ceremony. The marriage should be performed under a Tulasi plant. The epithets of Tulasi are 'Brahmani of Alambayan Gotra', great granddaughter of Vishvakarma, Prajapati's granddaughter, daughter of Ishvara. Sometimes Tulasi is also personified. She is described to be of black colour, lotus-eyed, smiling like a lotus bud, four-armed, decorated with ornaments and crown, white-clothed, sitting in *padmasana* and the consort of Vishnu. She is identical with the goddess Lakshmi.

Tulasi is worshipped like other gods and goddesses. She is believed to bless the devotees with plenty of riches, happiness, progeny and virtues. The women devotees are blessed with long life of the husband and children, prosperity, fortune and wealth. When a Hindu is in a dying stage Tulasi leaves are placed in his/her mouth believing that the person after death will go to heaven.

> A Vaishnavite deity takes a leaf of Tulasi with meals. The ceremonial dinner of a devotee Hindu begins with a Tulasi leaf. No Hindu meal is complete without one or two leaves of Tulasi. The Hindus take oaths also by Tulasi leaf and holy water in their hands. Tulasi is known as mother also. Hence, she is called 'Tulasi Mata' or 'Mother Tulasi'. Worshippers of Tulasi are also granted children, wealth, prosperity and happiness. To the devotees, the gate of heaven remains open. It is also said that the area about 3.2 km around Tulasi plant becomes pure. It is regarded that the holy Ganges is at its roots. Brahma resides on its branches and several other gods and goddesses live on its leaves. This plant is regarded to be the meeting point of heaven and earth. The Hindu devotees worship Tulasi for intellect, knowledge, for shelter in heaven also after death. Tulasi is worshipped by women folk for rich clothes, valuable ornaments, reserve of old stock of paddy, good and sound health of the husband till the end of her own long life. Virgins pray to Tulasi for getting worthy and suitable husbands. If anybody laughs at a devotee the former is cursed by Tulasi. It is mentioned in the *Ramayana* that Vibhishana had in his house a forest of Tulasi Plants.[34]

PIPAL (*FICUS RELIGIOSA*)

Pipal is one of the most important sacred trees held in high esteem not only in the mainstream of Indian religions but also in the religions of the tribal societies. Hindu saints and sages plant

---

34. T.C. Majupuria, *op. cit.*, p. 83.

Ashvattha (Pipal) and Banyan trees in and around their hermitage and practise penance under their shades. Among the sacred trees, Pipal and Banyan hold a very prominent position in the Hindu world. The Hindu trinity Brahma, Vishnu and Maheshvara are believed to live in them. These trees are immortal and imperishable. It is remarkable that the Bodhi tree never sheds leaves in summer and autumn and developes new leaves in time of Buddha's anniversary of *nirvana*. Buddha was enlightened sitting under this tree. So the tree is called tree of enlightenment. It is stated that after his Buddhahood Gautama moved to a Banyan tree. Its trunk represents connection between the invisible and the visible world, and its branches and roots represent an endeavour for perfection.

The wood of the tree is not used as fuel, as the Hindus believe that cutting of its branches will bring disaster. The tree is worshipped by Hindu women on the black moon day (Amavasya) for the general well-being of their family members. Water and milk are poured over its roots, sandal paste and vermilion are anointed on its trunk. Circumambulation for 108 times around the tree is considered to be a sacred act. On the occasion of marriage, sacred-thread ceremony and on other auspicious occasions the tree is propitiated. In some places barren women go round it naked with the intention of having a child. In some places it is a common custom to hang a pitcher full of water with a hole below to let the water drop on it roots. It is believed that the *grama-devata* or the ancestral spirits living in it are appeased. The *Bhagvad Gita* mentions that Vishnu appeared on the earth in the shape of a Pipal tree.

The tree has a very long life. The Bo tree in Ceylon believed to have been planted in 288 BC. bears testimony to the fact. The Bodhi tree in Bodh Gaya under which Buddha attained enlightenment is also very old. Of the many sacred trees recorded in the Puranas Pipal is one. This being the abode of Shiva, the snakes (Nagas) live in the tree. In some places justice is delivered under the tree. The judge sits under the tree along with people and the culprits. No one ventures to tell a lie to the judge. There is a common belief that Krishna stole the clothes of the milk-maids and kept these in the branches of a Pipal tree.

The Pipal tree provides cool shade to the passers-by. This tree is usually planted in the premises of Hindu temples and Buddhist complexes. A legendary account records the tree as the abode of Lakshmi. Lakshmi appeared from the churning of the milk ocean. The gods requested Vishnu to marry her. Lakshmi interrupted at the proposal that she would not marry before the marriage of her elder sister, Lakshmi of non-Aryan origin. Vishnu appreciated the idea and married the non-Aryan Lakshmi, who later on went away and took shelter on the Ashvasttha tree. Since then the Pipal tree became the abode of the non-Aryan Lakshmi.

Pipal is the eleventh Rudra and one of the Ashtavasus and it is like Narayan, the king of trees. Pipal is also stated to be the Vedic tree. There are several *shloka*s and *mantra*s in the *Rigveda*, *Yajurveda* and *Samaveda* referring to the tree. Thus this is a Vedic tree worshipped by the saints and sages. According to the Puranas if a married lady is predicted to become a widow she along with her husband and parents observes the Pipal fast to avert the disaster. The Tantra records that the virgin should plant a small Pipal tree and worship it continuously for one month. Then the marriage of the girl with the tree is solemnised on an auspicious day. This is done for the welfare of the family members, for the betterment of the house, for blissful conjugal life and for averting widowhood. Here the marriage ceremony is performed according to the Hindu marriage rites. The virgin and the pipal tree are tied ten times together with a thread. After some time the thread is removed and a mark is put on the forehead of the girl.

The tree has a great medicinal value. A paste is prepared from the bark of the tree and it is applied to any kind of inflammation. The juice is taken for relieving tooth pain and strengthening the gums. The juice is also very useful for any joint crack. The powder of the bark is applied for curing fistula. The fruit of the young leaves is used with water as a sure cure medicine for asthma.

BANYAN TREE (BARA – *FICUS BENGHALENSIS*)

The Banyan tree has been held in high esteem since the ancient times. It occupies a permanent position in the religious life of the people. It is worshipped both by Hindus and Buddhists. This tree is very famous for its unique feature of shooting down huge size branch roots which in course of time develop into strong new trunks covering a vast area. Its arial roots support the tree. It is also believed to be imperishable. The scriptures record that at the time of the deluge the Banyan tree was the only remnant on the globe. So it is designated as an eternal tree. The *Ramayana* of Valmiki refers to it as Akshay-bata (imperishable banyan). According to a popular story Satyavan cut a dried branch of the Banyan tree, and as a result he succumbed to death. His wife Savitri could restore the life of her husband by virtue of her single-minded devotion and propitiation to the banyan tree. In memory of this event the Banyan tree is worshipped on the occasion of Savitri Amavasya (15th day of the dark-moon fortnight in the month of Jyeshtha).

In some places the Hindus call it Vaivadha as people desiring vengeance against enemies propitiate the tree. The *Ramayana* relates that Rama in course of his *Vanavasa* lived in a Banyan grove. In ancient times the sages and saints lived under the Banyan tree and meditated for their whole life. The Vedic people in the early stage of their inroads into India conducted religious prayers and ceremonies under the Banyan trees. The Aryans held the tree very sacred. In the Buddhist texts the Banyan tree is depicted as the abode of Nagas. The snake goddess Manasa is also associated with several trees and plants including the Banyan.

One of the seals of the Indus valley civilisation depicts Lord Shiva under the Banyan tree. Possibly the Banyan tree in the pre-Vedic period was identified with Shiva. Lord Krishna is also a tenant of the Banyan tree. Lesser deities such as Yakshas, Kinnaras and Gandharvas also dwell on it. This supernaturalism is attached to it and it is regarded that the supernatural power of various potentialities guards over the interest, prosperity and destiny of the people. Sacred trees also protect individuals or the community from calamities like famine, drought, epidemic and other evils. The Hindus plant Bara, a Banyan tree, to have a seat beneath it for meditation to attain salvation through control over human action. The evergreen profuse leaves of the tree, abundance of fruits and its large growth have been symbolised with Lakshmi and Kubera. The tree is also believed to be the symbol of the goddess Shashthi, the deity of fertility. The goddess Shashthi is associated with children. On the occasion of her worship (in Jyeshtha) a branch of the *bata* is planted representing the goddess. She is worshipped on the 21st day after a child's birth.

The Banyan tree has great medicinal value. Infusion of the bark is used in diarrhoea, dysentery and diabetes. It reduces blood sugar. Latex or milky juice produced by the tree is used in treatment of dysentery, diarrhoea, lumbago, sores, ulcers, toothache, etc. The seeds of the Banyan tree have a narcotic effect, and so it is used as a tonic.

BEL OR WOOD APPLE (AEGLE MARMELOS)

Bel, as one of the ancient trees, is associated with different gods and goddesses of Hindu religion. Lord Shiva is closely associated with the Bel tree. The Bel in Sanskrit is Shriphala. The Hindus believe that the leaves of the Bel tree are offered to Shiva to cool and reduce the heat of his body caused by drinking poison coming out from the mouth of the snake at the time of the churning of the ocean.

During the funeral ceremony of some tribes a branch of the tree is placed on the roof to ward off evil spirits. Bel is worshipped by the people of all castes and sects who propitiate it as the incarnation of Shiva. Some clans of the Santal, Oraon, Birhor and Munda are associated with Bel and hence they worship it on ceremonial occasions. The Santal use Bel fruits in their Jitia festival observed by the women for the welfare of their children. The Nair of Kerala believe that Bel leaves and fruits increase the vitality of men and remove the virility of women.

The wood of the Bel tree is never used for fuel for fear of incurring the displeasure of Shiva. There are various stories and legends connected with the Bel tree. One such story relates that a hunter once having lost his way in the forest climbed a Bel tree at night being afraid of wild animals. Unknowingly he plucked its leaves and threw these down continuously throughout the night. There was a Shiva-linga (a stone symbolising fertility) under the tree. Shiva was satisfied at the gesture of the hunter and helped him to proceed in his mission next morning.

The Bel tree is closely associated with the manifestations of Shakti. The leaves and fruits are offered to Durga, Parvati, Saptamatrikas, etc. The married women worship the Bel tree for offsprings and for the well-being of their children and husbands. Virgins worship the tree with the intention of having handsome and loving husbands. Lakshmi, the consort of Vishnu, used to worship Shiva regularly with one thousand lotuses. One day the number of lotuses ran short by two. Lakshmi confronted a problem in her propitiation. Having no alternative to make up the number of one thousand she started cutting her breast with a sharp knife. When she cut one of her breasts Shiva appeared and stopped her to refrain from such an act. He was highly pleased with the devotion of Lakshmi. He blessed her that her cut breast would turn into the sacred fruit of the Bel tree. According to the *Agni Purana* Lakshmi was born in the shape of a cow and a Bel tree grew from its dung.

Every part of the Bel tree has medicinal value. The unripe or ripe Bel is used in treatment of chronic diarrhoea, dysentery and dyspepsia. The ripe fruit is very tasteful and prevents the disease of piles. The fresh juice of the leaves mixed with honey is given as a mixed laxative in times of fever and cold. It is very effective in asthma, constipation and jaundice. The great Ayurvedic physician Charaka advised the use of a cot made out of Bel wood for the labour room.

MANGO (*MANGIFERA INDICA*)

Mango is one of the most popular and useful trees in India. The Mango fruit is often expressed as the king of all fruits. The Mango flowers known as Manjari are believed to excite and improve the sexual urge. Twigs of Mango trees are used in almost all Hindu rituals. The sacrificial fire (*hotra*) is made out of its wood. The funeral pyre made of Mango wood is considered to be very sacred. In the sacred-thread ceremony of the brahmanas, the wooden seat on which the pupil sits is made of Mango wood. On ceremonial occasions the doors and entrance of the brahmana house is decorated with twigs and leaves of Mango. A Mango twig is always used on the *kalasha* (pitcher) placed in front of the deity. Among the five leaves included in the worship the Mango leaf is one. There are several mythological stories connected with Mango trees. Lord Krishna married Rukmini in the Madhubana full of Mango, Kadamba, Shami and Ashvattha trees.

A Mango tree is also considered a totemistic plant among the tribals. Destruction of such a plant by the clan members is a taboo. Among some tribes marriage with a Mango tree is a common practice. In some places the bride and the bridegroom are married to a Mango tree. Mango trees are planted invariably in the temple premises and at the outskirts of the village and under the mango tree the *grama-devata* (village deity) is installed. The tribals observe Ambanuakhia festival when the Mango fruits ripe. They eat mangoes and store their kernels for the lean season.

All parts of the Mango tree contain great medicinal value. The powder made from the tender leaves is used in case of diarrhoea and diabetes. A decoction prepared from the tender leaves with bark of castor oil plant is prescribed for cough, asthma and diseases of the chest. The smoke of the burning leaves of Mango are inhaled to cure throat diseases. The ash of the burnt leaf is used in case of burns. The leaves and soft branches are chewed to give tone to the gums. The bark is used as a tonic for mucous membranes. It is prescribed in case of disorders in bowels and the uterus and for dysentery, bleeding piles, haemorrhage in the lungs, diarrhoea, etc. The juice of the bark is also prescribed for gonorrhoea and as a remedy for diphtheria and other throat diseases. The dried flowers are used for diarrhoea, dysentery and several other diseases. The ripe fruit contains vitamin A, B, C, D and E. It is a laxative, astringent, stimulant and useful in haemorrhage of the uterus, lungs and intestines. Thus, on account of its multifarious utilities, the Mango tree occupies a significant place among India's trees.

LOTUS (PADMA – *NELUMBIUM NUCIFERA*)

Lotus is regarded as the most important flower in the mainstream of Indian religions as well as in folk religions. In a sense it has been enjoying an unparalleled popularity since earliest times. The Lotus finds profuse description in literature and art from the Vedic period. It is accepted as the emblem of beauty, wisdom and knowledge. In the *Rigveda* white and blue varieties of the Lotus are described. The *Atharvaveda* compares the Lotus flower with the human heart. In the *Rigveda* the blue variety of the Lotus is called Pushkara. The Ashvin Kumars, the twin gods of the dawn, became Pushkararaj by using blue Lotuses. In the *Brahmanas* the Lotus is associated with Prajapati, the creator of the universe. According to the *Taittiriya Samhita* the desire to evolve the universe saw a Lotus leaf coming out of the water. Then the universe came out on the thousand-petalled pure gold Lotus radiant like the sun. Lotus is thus believed to be the first product of the creative principle. The *Taittiriya Aranyaka* mentions that the universe in the beginning was fluid. Prajapati alone was created on a Lotus leaf. Similarly the *Mahabharata* describes Brahma to have sprung from the Lotus that grew out of the navel of Vishnu. In this Epic Brahma is described as Lotus-born or Abja-ja.

Lotus is closely associated with Lord Vishnu. Hence one of his epithets is Padmanabha (Lotus-navelled). The mythology describes that from the forehead of Vishnu sprang a Lotus from which was born his consort Shree or Padma or Lakshmi. Lakshmi has several epithets like Padmaja (Lotus-born), Padmasambhava (Lotus-born), Padmastha (standing on the Lotus), Padma-varna (Lotus-coloured), Padmakshi (Lotus-eyed), Padmini (Abounding lotuses), Padmamalini (decked with Lotus ornaments), etc. Lakshmi is depicted standing or seated with two Lotuses in her hands on the Lotus pedestal.

In the Tantric literature Kali is associated with the Lotus flower. Here Lotus is a symbol of eternal generation. In the Hindu, Buddhist and Jaina pantheons the Lotus seats and pedestal are universally adopted. All the gods and goddesses of these pantheons are depicted on Lotus pedestals. Lotus is one of the *ayudha*s of Vishnu. The image of the sun holds two fully blossomed Lotuses in his hands. The Lotus is also identified with the sun. This is due to the fact that the Lotus flowers blossom only after the sun rises and closes at sunset.

Lotus has great significance in Buddhism. The gods and goddesses of this pantheon are not only associated with the Lotus pedestals but also are beautified with Lotus flowers having stalks. The image of Buddha is shown sitting or standing on the Lotus pedestal. Manjushree, Maitreya and Avalokiteshvara are shown seated on Lotus pedestals holding Lotus stalks in one of their hands. Avalokiteshvara is very often known as Padmapani (Lotus in hand). The Bodhisattva is referred to by the Buddhist as *dharani om mani padme hum* (Oh jewel in the Lotus; Amen). The Buddhist goddesses Tara and Prajaparamita are associated with Lotus. Lotus is of different colours — red, pink, dark-coloured, blue-coloured and white lily. The plant is large and erect, grown in ponds and tanks. Its root is stout and crippling; leaves are round and standout high above the water. Almost all the parts of the plant have medicinal value. The juice of leaves is prescribed for diarrhoea. A paste of leaves with sandalwood is anointed on the body of a person when it is in high fever. The leaf stalks are used to cure headaches and young leaves are used in case of piles and leprosy. The flowers, their filaments and stalks are used as a cardiac tonic.

Conclusion

I have attempted in this paper to deal with the relationship of man with the forest in an ethno-historical perspective, touching upon different phases of culture from the prehistoric period down to the present day, in relation to plants and trees. Human attitudes towards the vegetation world from utilitarian, aesthetic and spiritual points of view deify certain plants and trees believed to possess supernatural powers.

Leaving aside the utilitarian and aesthetic aspects the nature of relation between man and forest is founded on the solid rock of religion. Everything was sanctified on the altar of religion. Archaeological and literary evidence and the oral traditions of the pre-literate societies in the form of legends, myths, stories and songs, corroborate the growth of human relationship with the forest over the ages.

Saints and sages in the attractive forest situation meditate peacefully and compose the great Epics, Puranas and Tantras which form the treasury of our knowledge. Forest is supplying food and shelter to animals and human beings. Trees and plants provide delicious and sweet fruits. Hence, these command respect from human beings. Forests are the abodes of deities and trees. It is also believed that trees are possessed by spirits both good and bad, which are the cause of diseases and disasters. Trees are regarded to bless fortune, destiny, progeny, etc., and to ward off evils. Therefore, this belief led to the worship of plants and trees by the Hindus.[35] The forest is indispensable in restricting soil erosion and preventing inundation caused by flood and checking up atmospheric pollution. Plants and trees are of great medicinal value. The traditional Ayurvedic practitioners and Tantrikas use medicinal herbs to successfully cure various diseases.

> Trees and plants with their herbage, flowers and fruits invoked intellectual or sensual perception. These natural things were beautiful and awful. Therefore, they became object of adoration and worship.[36]

James Fergusson has nicely highlighted the importance of trees

---

35. T.C. Majupuria, *op. cit.*, p.14.

36. *Ibid.*

and tree worship in his book *Tree and Serpent Worship*.

> There is such wondrous beauty in the external form of trees, and so welcome a shelter beneath their overarching bough, that we should not feel surprise that in early ages groves were considered as the fittest temples for the gods. There are also, it must be remembered, few things in nature so pleasing to the eye as the form or the colour of the flowers which adorn at seasons the whole vegetable kingdom and nothing so grateful to the palate of the rude man as the flavour of the fruits which trees afford. In addition to these were the multifarious uses to which their wood could always be applied. For buildings, for furniture, for implements of peace or war, or for ornament, it was indispensable. In ancient times it was from wood alone that man obtained that fire, which enabled him to cook his food, to warm his dwelling or to sacrifice to his gods. With all their poetry, and all their usefulness, we can hardly feel astonished that the primitive races of mankind should have considered trees as the choicest gift of the gods to men and should have believed that their spirits still delighted to dwell among their branches, or spoke oracles through the rustling of their leaves.[37]

On the basis of archaeological evidences he has indicated the prevalence of tree and serpent worship in Egypt, Judea, Mesopotamia, Africa, Persia, India and other countries.

On account of the above-noted factors and of various other things trees have been the subject of adoration around the globe. Although in the recent decades forest land has decreased substantially and tree species are in the process of extinction it is reasonable to pay attention to the past particularly on this aspect as the contemporary religious thoughts are based on the

---

37. J. Fergusson (1971): *Tree and Serpent Worship or Illustrations of Mythology and Art in India*, Indian Reprint, Oriental Publishers, Delhi, 4, p.1.

understanding of history. The ideas of the past are the determining factors in customs, beliefs, institutions, groupings and social forms.

References

Dalton, E.T. (1872): *Descriptive Ethnology of Bengal*, Calcutta.

Elwin, Verrier (1955): *The Religion of an Indian Tribe*, London.

Ernfield, U.R. (1952): *Kadar of Cochin*, Madras.

Fergusson, F. (1971): *Tree and Serpent Worship or Illustrations of Mythology and Art in India*, Indian Reprint, Oriental Publishers, Delhi.

Frazer, J.G. (1950): *The Golden Bough*, (Abridged edition), London.

Goldenweiser, A.A. (1910): Totemism, An Analytical Study, *Journal of American Folklore*, XXIII, No.88.

Gupta, S.S. (ed.) (1965): *Tree symbol worship in India. A new survey of a pattern of folk-religion*, Indian Publication, Calcutta.

Kurup, A.M. (1962): Glimpses of Life and Culture of the Tribes of Madhya Pradesh, *Vanyajati*, Vol.X, No.4

Majumdar, D.N. (1950): *The Affairs of a Tribe*, Lucknow.

Majumdar, D.N. and Madan, T.N. (1961): *An Introduction to Social Anthropology*, Bombay.

Majumdar, R.C. (1993): *Cultural Heritage of India*, Calcutta, Vol. IV.

Majumdar, R.C. (1993, Reprint): *Botany in Ancient and Medieval India*.

Majumdar, R.C. (ed.) (1980): *History and Culture of the Indian People*, Bombay, Vol.II.

Majupuria, T.C. (1988): *Religious and Useful Plants of Nepal and India*, Department of Medicinal Plants, Botanical Survey and Herbarium, Kathmandu.

Mukherjea, C. (1962): *The Santals*, Calcutta.

Roy, S.C. (1938): *Oraon Religion and Customs*, Calcutta.

Roy, S.C. (1915): *The Oraon of Chotanagpur*, Calcutta.

Sahay, K.N. (1956): *Malpaharia Village*, (Unpublished M.A. Thesis, Bihar University).

Sahay, K.N. (1965): *Tree cult in tribal culture, Tree symbol worship in India*, Calcutta.

11

Forests and Tribal Economy

Shasi Kant

Introduction

INDIA'S forests have played an integral role in sustaining its people over many millennia. In addition to an abundance of nutritional, medicinal and subsistence goods, wilderness areas have provided the environment for spiritual and cultural expression of the Indian people. In this historical relationship between forests and people, the tribal communities of India have a dominant place. The prolonged symbiotic relationship between the tribals and forests was the basis of their relative seclusion that is an expression of their independence from the outside world for their survival (Jena 1997). In India, there are about 260 tribal communities who constitute about 7.8 per cent of the total population of the country. For these communities, forests represent a god, a way of life, a culture, a home, an employer, and a manufacturing industry. Forests continued to be the cherished home for many generations of tribals. In times of extreme hardships like famines, forests are their last source of survival (Prasad and Jahagirdar 1993). Hence, tribal life has been totally dependent on forests. However, with the passage of time their peaceful co-existence with forests has been disturbed. The balance between these forest dwellers and the forests has tilted due to extraneous pressures that has made the existence of tribals in forests rather difficult. However, even today, forests play a very significant role in tribal economies.

At the macro level, a survey of the forest regions of Gujarat indicated that nearly 22 per cent of the children go to forests for collection of non-timber forest products (NTFP), such as tubers, leafy vegetables, bamboo shoots and other products (Prasad and Jahagirdar 1993). A Government of India (1984) report indicated that in Bastar District of Madhya Pradesh, the average household income from non-timber forest products is Rs. 1,500 against the total income of Rs. 1,750, and this income requires no initial inputs other than labour. However, the contribution of forests to the total income is only one dimension of the critical role of forests in tribal economy. Other dimensions of a developing or under-developed economy are income distribution, seasonal distribution of income, and rate of return on factor input of labour (Kant *et al.* 1996). Hence, to fully understand the role of forests in tribal economy, the role of returns from forests should be examined in all dimensions of tribal underdeveloped economy. In India, few village-level studies have examined the role of forests in some of these four dimensions.

Kant and Mehta (1993) studied the village economy of the tribal village Motisingloti in Gujarat. The village consists of 60 households of the Vasva tribe. In 1990, the total village income was Rs. 733,849 and per capita income was Rs. 1,835 that was fairly low as compared to the national per capita income of Rs. 3,284. The total contribution of forest based activities (both from direct monetary income and consumption quantified in monetary terms) was Rs. 566,837 representing about 77 per cent of the total income. The per capita income from forestry activities was Rs. 1,417. The direct monetary income from forestry activities (from sale of forest products) was Rs. 2,24,222 that is 30 per cent of their total income and 87 per cent of the total monetary income. The total number of man-days spent in the main activities pursued by the villagers for subsistence, excluding work on their own agriculture land, was 17,090 and 92 per cent of this was forest related. NTFP collection accounts for 72 per cent of the forest related work and 67 per cent of the total work. Weighted average[1] rate of return on labour (ROL) from the NTFP collection was Rs. 13.76 as compared to Rs. 22 from casual labour in the forest department and rupees five in agriculture fields. The activities of NTFP collection were spread over eight months from December to July. In this study, only marketable NTFPs were included, and non-marketable NTFPs that were consumed by people were excluded. The study also did not address the issue of income distribution.

Kant *et al.* (1996) studied 12 villages from the Jamboni range of Midnapore District of West Bengal. Of the 12 villages, one consists of only tribal people, nine of mixed population (tribal and non-tribal), and two of only non-tribal people. The weighted average annual household income of a non-tribal household was Rs. 18,237 and of a tribal household Rs.11,444. However, in tribal households, the contribution of non-timber forest products (including fuel and fodder) varies from Rs. 1,595 to Rs. 3,649 (with a weighted average of Rs. 2,461) while in non-tribal households, it varies from Rs. 581 to Rs. 3,930 with a weighted average of Rs. 2,548. Income from NTFPs increased the total household income by 30.12 per cent for tribals and only by 17.31 per cent for non-tribals. In the case of tribals, the weighted average contribution of NTFPs is 69.15 per cent of the agricultural contribution, and in one village it is as high as 139 per cent of the agricultural income. Even in the village where it is lowest, it is equal to nearly one-third of agricultural income. As far as income distribution is concerned, the Gini coefficient[2] for the

---

1. Average that incorporates the weights for every number. For example, if the household income of 2 households is Rs. 1,000 and 3 households is Rs. 3,000 each, the simple average income is Rs. 2,000 {(1,000+3,000)/2} and weighted average income is Rs. 2,200 {(1,000*2+3,000*3)/(2+3)}.

2. The Gini coefficient measures income distribution. In the case of every household having the same income, the Gini coefficient will be zero. If all income, for instance, is held by one household, the Gini coefiicient will be one. Hence, the value of the Gini coefficient being closer to zero means less inequality in income distribution.

tribal population reduced from 0.201 to 0.173 after inclusion of income from NTFPs. These twelve villages are under the Joint Forest Management programme of India, and, hence, forest rotation returns from timber harvesting will be distributed equally among all households. Therefore, returns from timber will not have a real affect on income distribution. On the seasonal distribution side, more than half of NTFPs collected are available for 10 or more months. This study did not incorporate the fourth dimension — return on labour.

Kant (1997) studied four villages from the West Midnapore forest division of West Bengal and three villages from the Vyara forest division of Gujarat. Two villages from West Bengal are of mixed population (tribal and non-tribal), and all other five villages (two from West Bengal and three from Gujarat) consist of tribal population. In West Bengal, the average NTFP's contribution ranges from Rs. 2,686 to Rs. 3,588 (weighted average Rs. 3,712) for the tribals, and Rs. 2,779 to Rs. 3,074 (weighted average Rs. 2,915) for the non-tribals. Approximately two-thirds of the total NTFPs contribution comes from fuel and fodder. The average contribution of all forest products is much higher for the tribal population (weighted average — Rs. 4,459, 44.5 per cent of total income) than for the non-tribal population (weighted average Rs. 3,597, 38 per cent of total income). In Gujarat, the weighted average of NTFP's contribution for the total population that is only tribal, is Rs. 2,388 which is much less than the average of Rs. 3,144 for West Bengal. In Gujarat, the contribution of NTFPs in one village is comparatively small as compared to two other villages. It is because the village has degraded Teak forest, and the villagers hardly get anything other than fuel and fodder from it. The forest contribution always has a positive impact on the reduction in the income inequality. Roles of contribution of fuel, fodder and other NTFPs on income distribution have been examined separately. The improvement in the Gini coefficient due to fuelwood contribution is generally higher than the fodder contribution, and similarly the improvement in the Gini coefficient due to other NTFP's contribution is higher than the improvement due to fuelwood or fodder contribution. The returns from timber are limited to one or two months, whereas the returns from fuel and fodder are spread over the whole year. The returns from other NTFPs are mainly concentrated during the summer season. This study also did not address the issue of returns on labour.

Hence, all these micro-level studies illustrate the important role of forests in the tribal economy. However, none of these studies have addressed all the four dimensions together. In this paper, a micro-level study from the State of Madhya Pradesh is presented that addresses all the four dimensions of tribal economy. The study also includes all NTFPs (marketed and non-marketed as well as marketable and non-marketable).

Firstly, sample villages and methods of data collection are described. Secondly, the methodology used for valuation of non-timber forest products is given. Thirdly, the results are presented and discussed and finally, the paper ends with some policy suggestions.

Sample villages and data collection

Five villages — Roshanpura, Dahira, Pura, Bharipura, and Kolikhera — from the Raisen district of Madhya Pradesh were selected for this study. In the discussions, these villages are referred to as MP-1, MP-2, MP-3, MP-4, and MP-5, respectively. The total number of households in MP-1 and MP-2 are 29 and 22 respectively; out of these 23 and 19 households are tribal households while the remaining households are non-tribal. The other three villages are purely tribal villages, and the number of households is MP-3 to MP–22, MP-4 to MP-9, and MP-5 to MP–21. Only the tribal households from MP-1 and MP-2 have been included in this paper.

The data collected can be grouped in two classes: (i) demographic data and data related to different sources of income, such as income from agriculture, forests, and other sources, for example, wage earning; and (ii) data related to the valuation of non-timber forest products. For the first group, annual data (1994) were collected. A format was designed for the collection of demographic and household income data. The format consists of five sections; (i) demographic data; (ii) income from agriculture; (iii) income from forests; (iv) income from other sources; (v) non-timber forest products data. Data in this format were collected from each household by interviewing one older member of the household, preferably the head of the household. This data collection was based on the assumption that people can recall one-year-old data. The data for the non-timber valuation were collected in group meetings using pen ended questions and Participatory Rural Appraisal methods, such as relative ranking.

Methodology for valuation of non-timber forest products

Non-timber forest products play an important role in the tribal economy, but many of these products are not traded in the market. Hence, no market value can be assigned to these products. However, these products are valuable to the tribals. Hence, the valuation of these products is critical to evaluate the role of forests in tribal economy. Non-market valuation requires the elicitation of perceived preferences. In the last three decades, numerous methods have been developed for assigning monetary value to non-marketable products. These methods to assign monetary value to goods are grouped into direct and indirect methods. Direct methods, known as contingent valuation (Mitchell and Carson 1989), use surveys or interviews to obtain individual valuation for given goods and services, and are based on the assumption that the respondent understands the good/service to be valued and its contribution to his/her economic life. The indirect methods rely on the indirect use of market information, and four indirect methods are the travel cost (Clawson and Knetsch 1966), the residual value (Nautiyal 1980), the hedonic price (Rosen 1974), and the household production function method (Muellbauer, 1974).

Kant (1997) developed a methodology to value non-marketable NTFPs. Kant categorised all NTFPs based on important features of NTFPs and suggested an appropriate direct or indirect method for valuation of each category of NTFPs. He grouped categories of NTFPs and methods of valuation mainly into five groups: (i) comparative valuation with respect to some product that is marketed and consumed at home for the products that are consumed in the forest; (ii) contingent valuation for non-marketable products which are collected occasionally; (iii) price of a substitute for the products that have a substitute; (iv) residual value method for products that are non-marketable with their end products being marketable; and (v) contingent valuation for the products for which neither a substitute is available nor the end product is marketable. Kant's methodology has been used for valuation of NTFPs in this study.

Economic contributions of forests

The contribution of forests to the four dimensions of tribal economy — the contribution of forests to the total income, impact of forest income on income distribution, seasonal distribution of forest income, and the rate of return on labour for collection and processing of forest products — are discussed next.

CONTRIBUTIONS OF FORESTS TO THE HOUSEHOLD INCOME

Tribals in these five villages are mainly dependent on agriculture and other sources of income that is mainly wage-earning as a labour (see Table 1). The returns from forests play only a supporting role. In MP-1 and MP-2, incomes from wage earning activities are more than half of the total income and are higher than that of agriculture. But, in the case of MP-3, agriculture contributes about two-third of the total income, and wage-earning activity about one-fourth. However, in MP-4 and MP-5, the contribution of wage-earning activity is minimal (less than 7 per cent), and agricultural income contribution is more than three-fourth of the total income. The contribution of forests varies from 3.6 per cent in village MP3 to 12.7 per cent in MP-1. The main contribution is from fuel and fodder. The contribution of income from fuel and fodder to the income from forests varies from 79 per cent (in MP-3) to 98.8 per cent (in MP-4) while the contribution of income form fuel and fodder to the total income varies from 3.6 per cent (MP-3) to 12.7 per cent (MP-1).

IMPACT OF FOREST INCOME ON THE INCOME DISTRIBUTION

In all the five villages, the forest contribution reduces income inequalities among village households. Gini coefficients are calculated for four cases: (i) income from all other sources except the contribution of forests (GC-1); (ii) income other than forests plus income from fuel and fodder (GC-2); (iii) income other than forest plus income from other non-timber forest products, excluding fuelwood and fodder (GC-3); and (iv) total income including total income from forests (GC-4). The calculated Gini coefficients are given in Table 2. The contribution of fuel and fodder improves the income distribution in all the five villages. In all cases, the contribution of other non-timber forest products also reduces income inequalities, and reduction in inequality due to other non-timber forest products is higher than that of the reduction due to fuel and fodder contribution. The reduction in income inequalities due to the total contribution of forests is obviously higher than the contribution of fuel and fodder and other non-timber forest products.

The percentage changes in the Gini coefficients due to the contribution of fuel and fodder, other non-timber forest products, and the total forest returns are also given in Table 2. The range of reduction due to the contribution of fuel and fodder is 0.2 per cent to 1.5 per cent only while due to other NTFPs it is 3.5 per cent to 6.7 per cent. The range of reduction in the Gini coefficient due to the total contribution of forests is from 4.2 per cent to 7.8 per cent. The improvement in the income distribution is highest (7.8 per cent reduction of the Gini coefficient) in the village that had the best income distribution (lowest Gini coefficient, village MP-1) before the inclusion of income from forests. However, the second best improvement (7 per cent reduction in the Gini coefficient) is for the village MP-5 that has the worst income distribution without forest income (Gini coefficient of 0.646). Hence, a general hypothesis that the relative

Table 1: Composition of average annual household income (in Rupees)

Source	*MP-1*		*MP-2*		*MP-3*		*MP-4*		*MP-5*	
	Rs.	%	*Rs.*	%	*Rs.*	%	*Rs.*	%	*Rs.*	%
Agriculture	4,125	27.8	10,362	39.5	29,155	67.0	36,927	88.3	24,011	83.3
OS	8,799	59.4	14,099	54.0	12,795	29.4	1,323	3.2	1,876	6.5
FF	1,588	10.7	1,466	5.6	1,245	2.9	3,522	8.4	2,844	9.9
ONTFP	0,302	2.0	0,229	0.9	0,331	0.7	0,44	0.1	0,73	0.3
Total Forest	1,890	12.7	1,695	6.5	1,576	3.6	3,566	8.5	2,917	10.2
Grand Total	14,814		26,120		43,486		41,816		28,804	

OS: Other Sources of Income; FF: Fuelwood and Fodder; ONTFP: Other Non-timber Forest Products

Table 2 Impact of forest income on income distribution

Gini	*MP-1*		*MP-2*		*MP-3*		*MP-4*		*MP-5*	
Coefficients	*GC*	*%C*	*GC*	*%C*	*GC*	*%C*	*GC*	*%C*	*GC*	*%C*
GC-1	0.391		0.430		0.597		0.465		0.646	
GC-2	0.385	1.5	0.427	0.6	0.592	0.8	0.464	0.2	0.644	0.3
GC-3	0.366	6.4	0.407	5.3	0.576	3.5	0.440	5.4	0.603	6.7
GC-4	0.361	7.8	0.404	6.0	0.572	4.2	0.439	5.6	0.601	7.0

GC: Gini Coefficient; GC-1: Gini coefficient for the total income except the contribution of forests; GC-2: Gini coefficient for income other than forests plus income from fuel and fodder; GC-3: Gini coefficient for income other than forest plus income from other NTFPs; GC-4: Gini coefficient for the total income including income from forests. %C: Percentage change in Gini coefficient as compared to GC-1.

Table 3: Income source-wise monthly distribution of average annual household income
(in per cent)

Month	MP-1				MP-2				MP-3			
	Inc. (Rs.)	*Agr. (%)*	*Oth. (%)*	*For. (%)*	*Inc. (Rs.)*	*Agr. (%)*	*Oth. (5)*	*For. (%)*	*Inc. (Rs.)*	*Agr. (%)*	*Oth. (%)*	*For (%)*
January	1152	0	88	12	1408	0	90	10	1558	0	92	8
February	1066	0	89	11	1311	0	91	9	1501	0	93	7
March	1029	0	89	11	1192	0	90	10	1307	0	92	8
April	921	0	88	12	1076	0	90	10	1164	0	92	8
May	1069	2	68	30	1186	0	75	25	1359	0	69	31
June	2207	61	5	34	7035	86	1	11	14285	94	5	1
July	747	0	87	13	900	0	89	11	944	0	90	10
August	760	0	88	12	925	0	90	10	920	0	90	10
September	814	0	88	12	966	0	89	11	893	0	89	11
October	993	0	89	11	1027	0	89	11	1126	0	91	9
November	1107	0	89	11	1201	0	90	10	1279	0	91	9
December	2949	61	35	4	7893	84	14	2	17150	92	7	1
Total	14814	28	59	13	26120	39	54	7	43486	67	29	4

Inc.: Total income Agr.: Contribution of agriculture. Oth.: Contribution of other sources. For.: Contribution of forests.

Table 3 (cont.) Income source-wise monthly distribution of average annual household income (in per cent)

	MP-4				MP-5			
Month	*Inc.*	*Agr.*	*Oth.*	*For.*	*Inc.*	*Agr.* (%)	*Oth.*	*For.*
	(Rs.)	*(%)*	*(%)*	*(%)*	*(Rs.)*	*(%)*		*(%)*
January	474	0	28	72	469	0	39	61
February	427	0	30	70	412	0	41	59
March	407	0	25	75	395	0	38	62
April	413	0	24	76	406	0	38	62
May	444	0	33	67	467	0	34	66
June	17749	98	0	2	11375	97	1	2
July	326	0	36	64	304	0	44	56
August	341	0	39	61	317	0	46	54
September	361	0	28	72	348	0	39	61
October	392	0	26	74	378	0	38	62
November	438	0	23	77	436	0	38	62
December	20044	98	0	2	13497	97	1	2
Total	41816	88	3	9	28804	83	7	10

improvement in the income distribution due to the contribution of forests will be proportional to the degree of inequality before inclusion of forest income, does not hold.

Seasonal distribution of returns from forests

An important aspect of returns from forests to rural economy is seasonal distribution. Contributions of different sources to the monthly and annual aggregate income are given in Table 3. In all the villages, the returns from agriculture are concentrated mainly in the two months of June and December. The returns from other sources of income and forests are almost uniformly distributed. Normally, fuelwood is collected as and when required; similarly, animals go to the forests for grazing almost throughout the year. Other non-timber forest products are collected in their season, but due to the variety of NTFPs, returns from either NTFPs are spread over seven months. In the case of MP-1 and MP-2, where agriculture is the second best source of income, the seasonal cushion provided by forest returns is not very important because the returns from wage labour, which provides highest income, are spread almost uniformly throughout the year. But, in the case of other three villages, the contribution of agriculture is more than two-thirds of the total income, and in the case of MP-4 and MP-5, the contribution of forests is higher than the contribution of other sources. Hence, in these two villages the returns from forests provide the seasonal cushion against the seasonal variation of returns from agriculture.

Rates of return on labour

The tribal people are involved in different production activities, and each activity has different levels of remuneration. The rate of return on labour (ROL), i.e., money earned per work-hour for each productive activity will provide a measure of economic efficiency of the activity. A comparative study of ROL from different NTFPs will help in comparing the economic efficiencies of collection and processing practices of different NTFPs; this information can be used to identify NTFPs that require an immediate support for improving collection and processing practices. A study of the ROL from different marketing channels for the same NTFP will help in developing marketing support strategies to local NTFP collectors. Details of ROL are given in Table 4.

The labour wage rate is Rs. 24 per day. Assuming an eight-hour working day, the hourly rate works out to be Rs. 3. Returns on labour from Tamarind (*Tamarindus indica*) varies from Rs. 3.75 to Rs. 2.81, and the ROL from Mahul/Mahua (*Madhuca*

Table 4: Rate of return on labour from NTFP collection

Product	*Raw Produce and units*	*End Product and units*	*Marketing Channel*	*Total time (hours)*	*Market Rate (Rs./unit)*	*Total Returns (Rs.)*	*Return/hour (Rs./hour)*
Tamarind	Raw 20 kg	Processed 15 kg	L.M.	20	5.00	75.00	3.75
			L.T.	18	4.00	60.00	3.33
			T.D.C.C.	16	3.50	45.00	2.81
Khajuri leaves	Leaves	20 brooms	V.	20	1.00	20.00	1.00
			L.A.	16	0.90	18.00	1.10
			L.T.	18	0.90	18.00	1.00
Siali leaves	Leaves	Leaf plates 3 bundles (each of 80 plates)	L.M.	13	4.00	12.00	0.92
			T.D.C.C.	9	3.50	10.50	1.16

Table 4: contd . . .

Product	*Raw Produce and units*	*End Product and units*	*Marketing Channel*	*Total time (hours)*	*Market Rate (Rs./unit)*	*Total Returns (Rs.)*	*Return/hour (Rs./hour)*
Kendu leaves	Leaves	2 Bundles 3200 leaves	L.A.	13	9.00	18.00	1.40
Mahul seed	Seed 10 kg	Oil 1 kg	V.	13	30.00	30.00	2.30
Mahul flower	Flower	Flower 10 kg	L.M.	15	5.00	50.00	3.33
Flower	Wine	6 bottles	V.	17	12.00	72.00	2.47
Sal seed	Seed 6 kg	Seed 4.5 kg	T.D.C.C.	8	1.75	7.87	0.98

Table 4: contd . . .

Product	*Raw Produce and units*	*End Product and units*	*Marketing Channel*	*Total time (hours)*	*Market Rate (Rs./unit)*	*Total Returns (Rs.)*	*Return/hour (Rs./hour)*
Sal leaves	Leaves	Leaf plates 2 bundles	L.M.	13	4.00	8.00	0.61
			L.A.	9	3.50	7.00	0.78
			T.D.C.C.	9	2.50	5.00	0.56
			T.D.C.C. Agent	9	2.00	4.00	0.44
		Leaf bowls 4 bundles	L.M.	14	3.00	12.00	0.86
			L.A.	10	2.50	10.00	1.00
Bamboo Products	Bamboo	Dala* 12 pieces	V.A.	22	5.00	60.00	2.73
		Roddar* 1 piece	V.A.	34	120.00	120.00	3.53
		Kula* 3 pieces	V.A.	8	6.00	9.00	1.00
		Bausia* 2 pieces	V.A.	4	4.00	5.00	1.25

Table 4: contd . . .

Product	*Raw Produce and units*	*End Product and units*	*Marketing Channel*	*Total time (hours)*	*Market Rate (Rs./unit)*	*Total Returns (Rs.)*	*Return/hour (Rs./hour)*
		Bhuja* 3 pieces	V.A.	9	6.00	9.00	1.00
		Jhuri* 5 pieces	V.A.	8	5.00	10.00	1.25
		Jhampi* 2 pieces	V.A.	8	12.00	12.00	1.50

Abbreviations:

L.A.: Local Agent

L.M.: Local Market

T.D.C.C.: Tribal Development Cooperative Corporation

L.T.: Local Trader

V.: Village

V.A.: Village Agent

* The following utensils are baskets made out of bamboo: ***Dala*** is used in rural marriages for displaying jewellery; ***Jumphi*** is used to carry sweets and other edible items from one relative to another, and also used in rural functions. ***Kula*** is used basically for vegetables, ***Bausia*** is used from carrying *pan* (betel-leaf and betel-nut mixture). *Bhuja* is multipurpose and used for grain, vegetables, etc. ***Jhuri*** is a broom with bamboo handle. ***Roddar*** is a grain container made out of bamboo.

indica) flowers is Rs. 3.33. Hence, returns from these products are greater than the wage rate. But, the ROL rate from other products is less than the wage rate with lowest returns from Sal (*Shorea robusta*) leaves. In general, the collection of leaves seems less remunerative when compared to other products. The ROL rate from Sal leaves varies from Rs. 0.44 to Rs. 0.78, from Khajuri (*Phoenix sylvestris*) leaves — Rs. 1.00 to Rs. 1.10, and from Siali (*Pueraria tuberosa*) leaves Rs. 0.92 to Rs. 1.16. Returns from the collection of Kendu leaves are Rs. 1.40 that is almost half the wage rate. Returns from Mahul/Mahua seeds are Rs. 2.30, but returns from Sal seeds are Rs. 0.98. Returns from bamboo products vary from Re. 1.00 to Rs. 3.53. Returns from the same product vary as a result of the variation in prices paid by different marketing channels. However, all bamboo products are sold to the village agent, and the variation in returns from bamboo is due to the production of different products from bamboo. ROL is the maximum for Roddar[3] which is a big item and production of only one piece takes about 34 hours. The noticeable and important feature of these returns is that, except for a few items (Tamarind, Mahul/Mahua flower, and Roddar), the returns on all other items are less than the wage rate.

Another interesting feature of this study is that the ROL varies with marketing channels. In the case of Tamarind, the ROL varies from Rs. 2.81 to Rs. 3.75, a variation of Re. 0.94; this variation is almost one-third of the wage rate. An interesting feature of the returns on Tamarind from different marketing channels is that the lowest returns are from TDCC (Tribal Development Cooperative Corporations) and the highest returns are from the local market. In the case of Sal leaves, the ROL varies from Re. 0.44 to Re. 0.78. Even though these returns from all the four marketing channels are very small as compared to the wage rate, but the variation in the ROL rate from different marketing channels is intriguing. Returns on Sal leaves from the

---

3. Roddar is a very big grain container made out of bamboo (about five to six feet high and two to three feet in diameter) and used for storing foodgrains.

local agent are 80 per cent higher than that of the returns from the TDCC agent. In the case of Sal leaves also, the ROL from the TDCC is smaller than returns from the local market or the local agent. In the case of Siali leaves, the ROL varies from Re. 0.92 to Re. 1.16, and returns from the TDCC are higher that the returns from the local market. Hence, even though one of the objectives of establishing the TDCC is to provide the local market support for NTFP collectors, but its pricing policy is unable to provide the maximum return on the NTFP collector's time.

Conclusions

In these five villages from Madhya Pradesh, the contribution of forests to the total income of tribals is not as high as it was in the case of tribals from Gujarat and West Bengal (Kant *et al.* 1996 and Kant 1997). However, other features of returns from forests — impact on the income distribution, seasonal distribution, and rate of return on labour — are similar to the previous studies. Hence, forests are not very important from the total income perspective, but they play a positive role in income distribution and hedging the seasonal variations of income from agriculture. The lower returns from forests are mainly due to the availability of other sources of income, such as agriculture and wage-earning, where the rates of return on labour are probably higher than that of returns from non-timber forest products collection. Hence, the tribals in these villages collect mainly fuelwood and fodder, and some other NTFPs mainly for their own consumption. The absence of *Tendu patta* (Kendu leaves — *Diospyros melanoxylon*) collection, which is the major non-timber forest product in Madhya Pradesh, is another factor for low contribution of forests in these villages. Hence, in general, it can be concluded that the share of forest returns to the total income of tribals will depend upon the presence of other sources of income that provide higher returns on labour. However, normally, the returns from forests will reduce income inequalities irrespective of the total returns from forests, and forest returns will provide the protection to the tribals against seasonal and annual uncertainties in income from agriculture. Due to presence of forests and traditional use of NTFPs by tribals, these people can fall back on returns from forests in the years of agriculture crop failures. But, the returns on labour are not attractive from the collection of the most of NTFPs anywhere in the country. Hence, for the economic welfare of tribal people, the Indian government and state governments need to focus their special attention on this issue.

One of the interesting features of the returns from NTFPs is that there are no competitive markets to these products at the local level. Competitive markets do exist for many products, but not at the local level, and tribal people often do not have access to these markets. In such a situation, the major part of the value addition goes to those who value through processing, or, in the case of locally processed products, the major portion goes to the agencies working between local producer and consumer. Therefore, there is good scope for increasing returns from NTFPs to local people by policy interventions designed to promote NTFP-based village and cottage industries, to develop local technologies for the use of NTFPs, and to enhance marketing facilities (Kant *et al.* 1996). The establishment of the TDCC was aimed to increase the returns to the tribals. However, its contribution has not changed the situation significantly. The central and state government need to design new policies and strategies to increase the returns to tribal people from NTFP collection.

References

Clawson, M., and J. Knetsch, 1966, *Economics of outdoor recreation. Resources for the Future*, The Johns Hopkins University Press, Baltimore.

Government of India, 1984, *Report of the Committee on Forestry Programmes for Alleviation of Poverty*.

Jena, Mihir, K., 1999, *Forest and tribal sacred complex; a comparison of three tribes of Orissa*, Contribution to this volume.

Kant, S., and N. G. Mehta, 1993, Forest based tribal economy; a case study of Motisingloti village. Forests, Trees, and People, 20, April 1993, pp. 34-9.

Kant, S., J. C. Nautiyal, and R. A. Berry, 1996, Forests and economic welfare, *Journal of Economic Studies*, 23(2), pp. 31-43.

Kant, S. 1997, Integration of biodiversity conservation in tropical forests and economic development of local communities, *Journal of Sustainable Forestry*, 4(1/2), pp. 33-61.

Mitchell, R.C. and R.T. Carson, 1989, *Using surveys to value public goods; the contingent valuation method. Resources for the Future*, Washington, DC.

Muellbauer, J., 1974, Household production theory, quality, and the hedonic technique, *American Economic Review*, 64, pp. 977-94.

Nautiyal, J. C., 1980, An anatomy of stumpage value, *Canadian Journal of Forestry Research*, 10, pp. 135-42.

Prasad, R. R., and M. P. Jahagirdar, 1993, *Tribal situation in forest villages*, Discovery Publishing House, New Delhi.

Rosen, S., 1974, Hedonic prices and implicit markets; product differentiation in pre-competition, *Journal of Political Economy*, 82, pp. 34-55.

12

Forest and Tribal Sacred Complex

A Comparison of Three Tribes of Orissa

Mihir Kumar Jena

The religious dimension of plants and forests

THE religious notions and perceptions of certain plant species of the Hill Kharias, inhabitants of Similipal, make Sal (*Shorea robusta*) a sacred species. Honey bees have a good interaction with the Sal population as they make hives on tall Sal trees and also forage on Sal flowers. Because of the religious beliefs associated with tall and old Sal trees, the tribals (Hill Kharia) usually do not cut them. Atharadeuli is a sacred place in the core area of Similipal Tiger Reserve and well-known among both tribals and non-tribals as a holy place. Around Atharadeuli there is Zaheera, a place with a number of old and tall Sal trees. Both the Atharadeuli and Zaheera area remain completely immune to human interference around the year except on festive occasions. Because of strong religious beliefs of the Hill Kharias, forests around their area are very much preserved and conserved. On the whole Hill Kharia give importance to Sal trees, the indigenous climax vegetation, and thus give an idea of conservation through their indigenous management systems.

Similarly the Kuttia Kondh attribute manifold significance to the forest. Their understanding of evolution of a forest is associated with their coming into being with reference to the Sapangada hole where they emerged from under the earth to the terra firma. According to their perception narrated in their myths and Gaani (songs), Asan (*Terminalia tomentosa*) and Siali (*Bauhinia vahlii*) are regarded as the most holy plants. They worship, love and respect these plants immensely on religious grounds. Shifting cultivation is the mainstay of the Kuttia Kondh. While going in for shifting cultivation, they are very conscious of the regeneration of plant species in the area under shifting cultivation. Since they respect the Dharni Penu (earth goddess) very much, they are conscious of not disturbing the plant species which Dharni Penu has created for them. In other words, the ecological importance of the forest is drawn from their practice of shifting cultivation. No doubt the shifting cultivation is a destructive dependence on forests, yet it is constructive and conservative in some aspects. By swiddening one land they preserve the gene pool of their indigenous subsistence crops and by fallowing on the other they preserve the genetic diversity. By their very attitude of leaving a specific forest area as cremation ground, they conserve certain plant species. Nobody is allowed to cut or destroy any plants growing in the area, lest they harm the sentiments of Dukeli Penu (main god of cremation ground) who dwells in such trees. Hence on the basis of their religious beliefs they respect these plants and indirectly help to preserve and conserve plant species.

The myth of the third tribe, the Saora, reveals that they were the people living inside the woods in forests. They love dense forests and prefer to live there. They have a myth that their gods prefer to live inside the deep dense forest provided with streams and mountains. Neem (*Azadirachta indica*) is the most important tree in their religious beliefs. Neem trees as dense forest along with various types of trees are worshipped by the Saora and hence they usually do not cut such plants. Thus those plant species have been preserved. In such dense forests in Minajhola many rare species of fauna including elephants safely roam without much human interference. The Saora's notion of preservation and conservation of forest is reflected in their strong religious beliefs and their ecological and economic relations with plants, trees and forests.

Historical and geographical setting

Nature worship is one of the most ancient traditions of India. Though nature broadly accounts for great many living and non-living components, yet in general folk understanding nature is known through forests consisting of types of herbs, shrubs, trees, climbers, twiners which grow and develop with various types of fauna living therein. Left to nature without any human interference forests can flourish into different dimensions. Nature worship in India is a very ancient tradition, described in the Epics of Hindu culture like *Ramayana* and *Mahabharata*, in the ancient antiquities of human civilisation — Vedas, Indian folk literatures, etc. The forest is the house of a great diversity of plants and trees, useful to human well-being. Next to the plants, animals are the secondary components of forests. It is through religious attributes, notions and sacred beliefs that patches of forests in India are well-preserved till today. The interrelations of man and forests have given birth to many myths and legends in which forests are perceived in many ways: biologically, economically, emotionally, ecologically and are associated in one way or the other with all these religious beliefs.

It is a fact that strong religious beliefs and fear of deities, presumed to be residing in trees and forests, have helped the preservation and conservation of the natural habitat. Utilisation of forests as well as its preservation and conservation date back to the pre-agricultural period when man depended on animal hunting and food gathering in the forest for his existence.

Dravidians and Aryans are said to have lived in rapport with nature utilising forests and forest wealth unrestrictedly for centuries. References to forests and forest wealth in India have been recorded during Alexander's invasion. In 300 BC. Chandragupta Maurya had appointed a Superintendent of Forest whose task was the conservation and protection of forests.

Over the years conservation of forests and biodiversity in India has been marked by a continuous gap between aspiration and efforts and between preaching and practising. To earn religious merit and to ensure rational prosperity for his family, a man should protect the forest and trees. These are the 4000-year-old injunctions noted in the chanting of the *Agni Purana*. Every man should plant a tree in a five year's interval, preached Gautam Buddha 2500 years ago. The injunctions in the scriptures and the preaching of the saints made people venerate trees but did not lead them to protect and propagate trees and to conserve forests. In later days the conservation of plants and trees, forests, animals became a part of the eastern culture. The prolonged symbiotic relationship between the tribals and forests constituted the foundation of tribal independence in the forest. The mutual interdependence was the basis of their relative seclusion which was an expression of their independence from the outside world for their survival. However, the conservation and management of forests on a scientific basis were first introduced in India by the British administration.

Before the practice of scientific forestry in India, it were the people who preserved and conserved the biodiversity and forest wealth in their indigenous management systems. With the impact of modernisation, industrial development, overexploitation of forest resources for running industries and so many other factors in the name of economic development the symbiotic relationship of man and forest broke. Enormous quantities of forest patches were denuded to meet the demands of industries. The situation came to an alarming position. Deforestation did not happen overnight. Various types of environmental hazards resulted from deforestation. This developed global consciousness about the obvious environmental hazards. Scientific management practices for preservation and conservation of forests became an obvious necessity. It has long been felt that the varieties of organisms, vegetation and ecosystems cannot be safeguarded satisfactorily if they are not allowed to grow in protected and suitable areas. It became thus necessary to reserve the forests in the name of biosphere reserves considering the relationship between flora and fauna, indigenous population and environment. These reserves often reflect centuries of human experience and can provide information of immense value in improving the productivity and sustainability of ecosystems through conservation practices.[1]

Who are the indigenous people? The human evolutionary myths give an idea that prehistoric man was born in forests, and was reared in the lap of forests. Prehistoric man drew his subsistence and ensured his sustenance from natural forests. In the present scenario when the conservation of biodiversity has proved to be 'a million dollar question', interdisciplinary studies are being organised to evaluate the forest management systems by indigenous people. Here we regard the tribal people as the indigenous people, who since ages have been living amidst forests and are regarded as forest dwelling communities. They are also the forest dependant people as they derive the three prime needs of food, clothing and shelter from forests. Their manifold attachment to forests has created many myths and legends, ideas and perceptions, values and mores in which an intricate link of their society, religion and culture with forests is clearly exhibited.

In my quest for tribal notions on preservation and conservation of forests, I met three primitive tribes of three different regions of Orissa. They are the Hill Kharia people living in the vicinity of Similipal biosphere reserve, better known as Similipal Tiger Reserve, the Kuttia Kondh of Belgarh area of undivided Phulbani District and the Lanjia Saora of Puttasingi area of undivided Koraput District. All three tribes show unique cultural patterns and exhibit primitivity in their livelihood patterns, methods of food collection, various methods of extracting resources from forests, etc. In their religious framework forest is greatly valued, and various parts of these forests are given religious importance. Well known is the fact that in the tribal myths and legends forest is considered as a foster mother of the tribal people. Their indigenous and traditional knowledge about the effective utilisation of forests and exploitation of forest

1. Patnaik & Jena, 1995:9.

produce is embedded in their myths, legends, work traditions and social processes. Their knowledge in the management of forests and notion of preservation and conservation is embedded in their oral traditions and work patterns both theoretically and practically. The considerable wisdom of the tribal people on this particular aspect is transmitted from the predecessors to the successors either through oral traditions or through work traditions.

However, the intimate relationship between the forest dwellers and their green neighbours, plants, is well known. Primitive man depended on forests for all requirements and in turn protected forests by conserving sacred groves and sacred species and observing taboos in the use of certain species. In ancient times, due to a thinner population and very limited needs of people even shifting cultivation did not cause irreparable damage to forests and regeneration compensated removal of vegetation.[2] The tribal peoples' conservation practices of forests or trees is based upon religious and mythological grounds.

In this context I take the opportunity to elaborate the tribal religious beliefs and perceptions associated with certain forests under the heading 'Forests and tribal sacred complex'. It is through their ideas and strong conviction that preservation and conservation of forest could be maintained. I shall deal with the forests and tribal sacred complex in three sections, firstly focusing on the Hill Kharia, secondly on the Kuttia Kondh and thirdly on the Saora.

Hill Kharia

For my study amongst the Hill Kharia, I concentrated on the villages Jenabil and Guruguria, the former being in the core area of Similipal sanctuary and the latter in the buffer zone. Hill Kharia are the original inhabitants of Similipal hills. Inside the Similipal biosphere reserve and in its outskirts dwell other tribal groups like the Bathudi, Ho, Santal, Mankidia, Munda, Bhumij, etc. Of

2. Jain & Hajra, 1979.

them the Hill Kharia are adept in collection of minor forest produce, the chief one being honey.[3]

Geographically, Similipal is an oval-shaped highland. It is the meeting ground of northern and southern flora of the country. The Project Tiger launched in 1973 caused India's spectacular success involving resource conservation. The forest is a north tropical moist deciduous forest with Sal (*Shorea robusta*) as the dominant vegetation. Besides Sal, Asan (*Terminalia tomentosa*), Piasal (*Pterocarpus marsupium*), Kurum (*Adina cordifolia*), Kusum (*Schleichera trijuga*), Kendu (*Diospyros melanoxylon*), Dhaura (*Anogeissus latifolia*), Jamun (*Syzygium cuminii*), Bandhan (*Ougeinia dalbergioides*) are the major timber species. Atharadeuli is a sacred place in the core area of Similipal Tiger Reserve. It is well-known among the tribal and non-tribal people as a holy place. On account of its relative inaccessibility, people attribute special significance to those pilgrims who are able to visit the holy place situated on the peak of a hill. Atharadeuli is very well recognised as one of the Mahu Bhandars (honey reservoir) in Similipal forest since long. Other Mahu Bhandars are prefixed by the name of the place of their location. The Kharia people strongly hold their opinion that Similipal was famous for honey. Now, Atharadeuli gets its importance as a honey reservoir.

The Hill Kharia believe that one's desire would be fulfilled if worship and sacrifice is offered to the god residing there. At the foothill there is a ruined temple which still stands as a proof of the presence of a god in Similipal in the ancient past. According to another local account, Yogi Mahanta Maharaj had built the temple there. The ruined temple, called 'Manav Deula', is surrounded by a number of trees. A Sal (*Shorea robusta*) and an Asan tree (*Terminalia tomentosa*) are found close to each other and they guard the temple where many gods exist. Besides the main god 'Ishvar' inside the ruined temple, gods under the trees are also worshipped. A little away from this place there is a pond where one can take a bath before entering the temple but

3. Patnaik & Jena, 1995:10.

no one goes near it, as it is believed that it devours living beings, i.e., men and animals. This pond is named 'Meghanad Pokhari' and it is believed that it leads to hell.

The origin of Atharadeuli temple 'Manav Deula' is described in the following story. Once after praying at the main temple of Atharadeuli the priest was returning to his house when he suddenly remembered that he had left his water jug inside the temple. He retraced his steps and found the goddess enjoying the food he had offered to her during worship. In a fit of anger the goddess threw the jug and cursed that no one could pray in that place any more, because a deity never should be seen at the time of taking the offerings. After the priest begged her forgiveness the goddess relented and told the priest to build the temple where the jug had fallen. Thus the temple Atharadeuli, i.e., 'Manav Deula' was built. Yet another myth, a rather long story linking Lord Shiva with Atharadeuli, exists, which cannot be told here as it is not within the scope of this paper. People fear that if any house is built within half a mile radius of the temple, evil spirits harm the members of the house. A case which is usually highlighted to state this fact is the following: a forest guard — Madhu — built a house near the temple, but as soon as the house was completed elephants came and destroyed it. The Asan tree and Sal tree that are found outside the temple are regarded as the abode of servants of Lord Shiva. Sal is very important in the religious beliefs of Hill Kharia. An ethnographic account of the Sal tree as perceived by Hill Kharia may be given here. Oral descriptions of Sal — the indigenous and predominant vegetation of the Similipal forest — emphasises the sanctity of the tree. It is sacred for the following reasons:

1. Sal is naturally germinating. The seeds mature and germinate while still attached to the mother tree. When they fall on the ground they immediately germinate.

2. Sal is the dominant tree in the Zaheera (sacred grove) where the tribals offer worship to satisfy the clan gods. The wood is used for the sacred fire. Leaves are used to

make leaf plates and bowls to serve food to gods. Sal resin is burnt to drive away mosquitoes. Therefore the gods like the tree.

3. Sal wood is regarded as 'Daru', meaning god.

4. The honey bees make hives on tall Sal trees.

5. Sal seeds are a lean season food for the Hill Kharia. The seeds are also an important minor forest produce.

6. It is a primary host plant of silk worms.

The Sal trees add importance to the Atharadeuli as a Mahu Bhandar (honey reservoir). The Hill Kharia's perception about the contribution of Sal to honey production is given by Patnaik and Jena[4]:

> The Hill Kharia are of opinion that honey bee is dominant among fauna with Sal (*Shorea robusta*) as the dominant flora. Honey bees have a good interaction with Sal population as they make hives on tall Sal trees and also forage upon Sal flowers.

In the Atharadeuli there is a huge population of very old and tall Sal trees from the branches of which many honey-combs are seen hanging. Because of the religious beliefs associated with tall and old Sal trees the people usually do not cut them. Again there is a number of old and tall Sal groves, regarded as Zaheera found around Atharadeuli. The Zaheera and Atharadeuli areas remain completely immune to human interference all round the year except on some festival days.

The forest around Atharadeuli is therefore very much preserved and conserved based on the religious beliefs of the Hill Kharia. My informant Sukra once told me that the floristic diversity of Atharadeuli area is so vast that one would never fail to get a desired plant species from that area.

---

4. Patnaik & Jena, 1995:11.

Kuttia Kondh

In this section I discuss the sacred complex of the Kuttia Kondh in relation to forests. The Kuttia Kondh are a primitive tribal subgroup of the Dravidian Kondh tribes of Orissa. They speak the 'Kui' language and live in the hills of Orissa. The Kuttia Kondh are forest-dwelling tribal communities. Their main source of livelihood comes from shifting cultivation, occasional hunting and minor forest produce gathering. The Kondh are well-known for their heinous human sacrifices. The pattern of their association with forests is highly reflected in their religion, economy, social customs and traditions, emotions and aesthetic values. My findings are gathered from the Kuttia Kondh of Burlubaru village.

The forest in and around Belgarh area falls under the northern tropical moist deciduous zone. Sal being the predominant species in pure form is distributed from the lowest latitude to the top of the hills. No doubt, the best crop is noticed in the valleys having deep loamy soil. Common associates of Sal in the region are *Terminalia tomentosa*, *Anogeissus latifolia*, *Lagerstroemia parviflora*, *Pterocarpus marsupium*, *Albizzia chinensis*, *Toona arborea*, *Xylia xylocarpa*, *Careya arborea*, *Bridelia retusa*, *Dillenia pentagyna*, *Buchanania Lanzan*, etc.

The Kuttia Kondh attribute manyfold religious significance to the forests. They know the forest by various clearly-defined terms. They are Kambani, i.e., canopy forests having large trees, Bati, i.e., the forest in general or jungle, Umda, i.e., patches of forest amidst a plateau or barren land, Tuleni, i.e., the patches of forest where dead bodies are cremated. The Kuttia Kondh perceive the evolution of a forest in connection with their own coming into being. They understand a forest with reference to the hole of 'Sapangada' from where all living beings emerge. The Kui language is deeply anchored in the forest landscape. Their oral tradition like the Kui Gaani (song) illustrates their coming into being along with plants and animals. Sapangada is a plain land circumscribed by many hills located in the Guma Panchayat, near the Kuttia village Kranja. The Kuttia Kondh myth states that once they came out of the earth through the hole of

Sapangada. Kuttia Kondh believe that Nerandali, the supreme lady, had given birth to them. This is the commonly accepted myth among this tribe. Some people hold that people were born under Dimbuli hill. But as they could not find any open space to come out to earth, they finally reached under Sapangada where a porcupine helped Nerandali in making a channel connecting the underearth to the earth, where they are living now. According to their perception the entire Sapangada area is divided into two main parts, viz. (1) Sapangada (birth hole) which was surrounded by Satikambani (born forest) and a Mardimara (Asan tree — *Terminalia tomentosa*), which is still standing on the border line of the Sapangada hole. Similarly another tree — Siali (*Bauhinia vahlii*) was also found to be there in another portion of the border line of the hole. These two trees hold a significant identity in their myth. The myth reveals that immediately after the entire population emerged out of the hole of Sapangada, a cow followed them. The people who came out first could pull out the cow by tying it with a rope made out of Siali (*Bauhinia vahlii*) bark. After the cow came out, they tied the rope to a plant. The Kuttia Kondh believe that this plant has then turned into an Asan tree (*Terminalia tomentosa*). Henceforward the Asan tree became their most precious and religious plant. The myth narrates that after coming out of the hole, the cow attempted to kill and eat all the people who had emerged from earth earlier. So Udurengad, a mythical hero, broke all its teeth. Blood oozed out from the cow's mouth and turned into several white ant hillocks.

According to the present priest of Sapangada, named Badu Pujari, son of Preeti Pujari of Village Kranja, the entire population of the people who came out from Sapangada were divided in two main categories, Rajianga and Kuenga. And these two broad divisions descended from two Kondh brothers named Mininguera and Dambinguera who were born in Raniadu. Mininguera was older than Dambinguera and their caste was then Kondh. Both brothers challenged each other to determine who was more powerful. For that purpose an elephant was kept in the middle of a fence of seven bamboo pieces. Dambinguera crossed the fence and the elephant on one leap. Mininguera failed to do so.

Thus, Dambinguera was declared powerful and became the king. He made his family known as Rajianga in which all other people except the Kuttia Kondh were included. Mininguera's family got the name Kuenga in which all the Kuttia Kondhs were included.

The Kuenga were then divided into many lineages. Many stories are associated with Nandruka and Timaka. The Nandrukas came out first from Sapangada. They had giant-like bodies with very long ears and large eyes. Nerandali was upset on meeting the Nandrukas. Such huge people were beyond her desire and imagination. So out of anger she cursed the Nandrukas that she would never provide anything for them and they would survive by eating their own body parts. Nerandali did not care to accept any type of offerings given by Nandrukas. Ultimately she drove away Nandrukas from Sapangada. The next to be born were the Timaka. But Nerandali did not feel like giving birth to the Timakas in the same manner as she had done for the Nandrukas. So the Timakas went out in search of a place to come out and finally reached Knehelpanga and Damanpanga where they were born from bamboo shoots (Terka). Nerandali felt like accepting offers from the Timakas. So the Timakas were valued as priests who worshipped at Sapangada so that the offerings could satisfy Nerandali. So Sapangada at present is being worshipped by the Timaka at an interval of five to 15 years. Group worship by the Timaka are usually organised at seven years' interval. The Timaka were first given crop seeds by Nerandali. The Timakas were later on allowed to distribute seeds and fields among the Kuttia people. The first distribution of seeds was done at Suna Sella, a hill at the eastern boundary of Sapangada near the village Pedem.

Paddy crops were given to the Rajianga people to cultivate. The other crops in mixed varieties like Kueri (*Panicum miliare*), Pokanga (*Zea mays*), Kabanga (*Ricinus communis*), Kating (*Vigna sps.*), Kredanga (*Phaseolus mungo*), etc., were given to the Kuenga people. Further, to solve the distribution of the land, the Rajianga were given the first priority and so they were allowed to cultivate plain land as paddy can hardly be cultivated on hill slopes. The Kuenga were allowed to cultivate in the Soru Nede region which is recently known as Nella. Further, the hills and slopes were distributed among the heads of different lineages; Mininguera was supervising the distribution process. The people of Damanaka distributed the lands to the Timakas. People now believe that the hills, where Dharni Vali is found, was definitely inhabited and cultivated by their forefathers.

The Bicha Supinere festival celebrated in the forest gives importance to a hill named Suna Sella. The festival is organised often after field preparations. The seeds are worshipped there with a sacrifice. This is done to beg a god's excuse for cutting down the forest for sowing the seeds. People have the apprehension that unless they do such offerings the gods may create excessive rains to destroy the seeds.

The next place of importance in Sapangada is the Mila Mitpen Vali where there was a pile of stones in ancient days. Some statues were also found there recently. Presently two stones represent the Mila Mitpen Vali, one larger stone being regarded as the mother in whose cradle there is a smaller stone regarded as the daughter. In ancient days a bath by the people in that locality was considered sacred. It was considered more auspicious if the children took their bath. Mothers used to take their children to bathe there with turmeric paste and water. Then they used to worship Sapangada Penu (God). The Sapangada Pujari (priest) presided over the invocation. 'Mila Mitpen Vali' means 'The Bathing Stone of Children'.

Three Dharni stones are found within a radius of two metres at the centre of Sati Kambani. The Dharni Vali stone which faces the East is the most important one. A stone is established as the Pujari's seat. Just after the Dharni Vali there are two spots which are used as poles for sacrifice. In ancient days human sacrifice was executed at those posts. After the sacrifice the dead bodies were buried under a Sal tree, which was at an angle of about 60° to the left of the Pujari seat. There are a number of stones at the base of the tree till today. Dharni stones facing south and west are also worshipped along with the Dharni stones facing the east.

Sal tree and Asan tree are given equal importance here from the religious and mythical points of view. The Kuttia Kondh perceive Sal in a more or less similar way as the Hill Kharia. Adding to that, Sal wood is used to make sacrificial posts, wooden posts representing deities in their material culture. Karumunda is the name of the wooden post used for buffalo sacrifices. In ancient times human sacrifice was given by tying the victim to this pole during Meriah festival. In some areas people use the pole of Asan tree (*Terminalia tomentosa*) instead of Sal pole as the Meriah pole. Their association with both the trees is based upon their mythical importance.

The ecological importance of the forest is drawn from their practice of shifting cultivation. The Kuttia Kondh are shifting cultivators since ages. Shifting cultivation is also known as swiddening or slash-and-burn cultivation. They utilise their indigenous knowledge on biological resources. In the swiddening practice they leave the patches of land for long fallow so that the cleared patch of the forest can regenerate. From the myth of their emerging out from Sapangada they are aware about shifting cultivation. They believe that the fertility of the soil is indicated by the vegetative regeneration of the cut trees and plants. Conversely, fertility is the indicator of vegetation. From this they believe that when the fertility of a land decreases to a great extent, a crisis sets in the improper growth of vegetation and so the yield is minimised. Besides, they think that the entire forest landscapes are owned by one of their ancestral spirits named Damanaka who is the sole authority of fertility and vegetative regeneration. Hence, before cutting a new patch of forest they ask permission from Damanaka through religious performances. They are aware that continuous cultivation on a land for years would decrease its fertility and that it is regained by long fallow periods as a result of which a lush green vegetation will regenerate indicating that the fertility has returned.

Their conservation practice of plants and trees is well-marked in their maintenance of Tuleni (forest left for cremation of dead bodies). Collection of firewood, cutting of trees for any purpose is totally prohibited in these areas. Even resin, mushrooms, tubers, herbs and medicinal plants are not collected from such forest patches. According to their belief any disturbance of the vegetation is a violation against the will of ancestral spirits, who otherwise cause diseases. Among different species of plants Sal is given a special status in Tuleni. For people believe that the Dukeli Penu (main god of Tuleni) dwells in such trees. In exceptional cases, the Kuttia Kondh medicine man, if permitted by Dukeli Penu, can collect and use certain herbs and roots for medicinal purposes.

Saora

The Saora (known by various names such as Savara, Saur, Sora, etc.) constitute one of the most primitive community of India showing racial affinity to the proto-Australoid group. Linguistically they are akin to the Mundari branch of Austric language. In the words of Verrier Elwin[5] "The Saoras who give the impression of being rather matter of fact and prosaic are surprisingly picturesque and metaphorical in their speech." The Saora trace their origin from Kureitung — a hollow trunk of a tree. From the myth of their origin they regard themselves as Sora. It is derived from two words, so (hidden) and *ara* (wood/tree). From this etymological meaning of 'Sora' the Saora pose as people living inside forests. They depend on forests in many ways to derive their livelihood.

In the Saora world-view, known in their language as 'Saora Jagat Katha', there are descriptions of the relationship of the first man with forests. The Saora believe Kittung and Radab to be their first ancestors. The Saora myth speaks volumes about Kittung. But the Shudha Saora people, a branch of Saora, take the myth some steps ahead from Kittung to Madusa (Madu: dead body; Sum for Sonum, spirit). Kittung died in the forest, killed by accident by a Saora man. Before Kittung came to life again as Madusa, this Saora man tried to cremate the dead

5. Verrier Elwin, 1955: 49.

body. He collected dry Bel (*Aegle marmelos*), Chandan (*Santalum album*) and Neem (*Azadirachta indica*) wood to burn the dead body. But the dead body of Kittung got life again and told the Saora man to keep him in a place unnoticed by human beings. Kittung's specific criteria for choosing a suitable place for his comfortable living was as follows: The place should be situated completely solitary in a dense forest of Neem (*Azadirachta indica*) groves. There should be a permanent source of water where a god was expected to take his bath and drink. The water in the nearby hill could keep the place cool during summer. A number of Siali (*Bauhinia vahlii*) climbers must surround the place to give a cool breeze. There must be bamboo bushes around the place, because the bamboos could create sparkling fire to warm during winter. More so the god wanted that the Saora man should keep the matter confidential. He cautioned the Saora that if the truth is unravelled, the god would leave the place for ever. As long as he stays in his secret shrine, the Soara should offer him fruits collected from the forests. The god lived in a Neem tree. A splendid glow in the Neem tree indicated the god's arrival at the site. The god warned that as soon as he approached the tree the Saora should be ready with offerings.

The Saora man, however, was able to find a place according to the god's design. The place is near Minanjhola. (Maina-Joli = rivulet or canal). The geographical location of the place is to the South of Kutra Garh — a place near Muniguda and is located somewhere on the northern border of Minanjhola. From Minanjhola three rivers emerged known as Urangjal, Furfurda and Runkuda, literal meanings of which are bamboo canal, swelling water and rice water respectively. The god liked the place. The fruits like Mango (*Mangifera indica*), Jackfruit (*Artocarpus heterophyllus*), Guava (*Psidium guajava*), Banana (*Musa paradisiaca*), etc. were offered by the Savar to the god. The god lived in a Neem tree. The next part of the myth describes how the god was stolen from that place by King Indradyumna of Puri. At Puri the god was established as Lord Jagannath.

The Saora people know the place Minanjhola as the holy shrine of Madusum who is the incarnation of their great grand ancestor Kittung. The non-tribal people know the place as a Shakti-sthala, seat of power of Lord Shiva and as a shrine of Lord Shiva and his wife Parvati. A huge gathering of tribal and non-tribal people is observed here on the day of Maha Shivaratri festival. The Saora regard the place Minanjhola as their great sacred complex, for the Jagannath cult is associated with the place. An analysis of the floristic composition in the area shows enormous bamboo bushes, very tall Sal trees and other species of the Sal community. Great numbers of Neem (*Azadirachta indica*) trees fill the interspaces of bamboo and the other associated species of Sal community. Numerous fruit-bearing trees like Mango, Jackfruit, Guava, Kendu (*Diospyros melanoxylon*) and Bel (*Aegle marmelos*) are found in the place.

The forest falls under the north tropical dry deciduous type with Sal as the dominant flora. The place is surrounded from all sides by perennial water sources. The deep and dense floristic composition covers the place and the place looks dark even during day time. The Saora mention that the forest caters to a number of elephants whom they believe to be the vehicle of god. The Saora people have a strong conviction that their god (Jagannath) cannot live away from them for long. One day he will definitely return to his old shrine leaving the decorated Puri Temple. The Saora people do not interfere with the vegetation of the place. They hold that they have preserved all sorts of fruit-bearing trees to cater to the needs of the god and also many a number of Neem trees. The god may choose any one to live in. All the things which the god desired earlier are still preserved, untouched and without any human interference. The elephant population is growing at the spot because they get enough food from wild banana plants, young bamboo shoots (in the rainy season), fruit-bearing trees and evergreen vegetation. Few tigers and some leopards are also found in the forest. The forest remains immune to human interference for two main reasons: first, because the place is regarded as the shrine of Madusum and secondly, one would not like to take the risk of being trampled under elephant's feet. Both things arouse fear in the people owing to which the people hardly gather forests produce from the spot. Some people normally do not cut Neem trees, as they believe that the god lives in Neem trees. They also do not interfere with Sal trees for the belief is that the god lives in the tree for dust resin and for their height that provides shade to the whole place and keeps the place cool.

Interference in such forest to fetch minor forest produce for economy in terms of cash is very much limited. Saora people go there mainly for bamboo shoots in the rainy season or for fruits during their ripening season. Sometimes they collect bamboo from the bushes located in the outskirts of the area. The Saora, however, could not restrict the non-tribal people from exploiting minor forest produce like leaves of Sal and Siali to make leaf cups and plates, dust resin, Mahua flower, fruits and roots, etc. Nevertheless commercial exploitation by local people is very negligible. In their economic relationship with the forest, the Saora believe that the god does not get angry if they collect food materials from the place. Their interference is limited to certain food materials only, therefore it is very casual. The Saora thus perceive the forest as a great reservoir of flora and fauna. Their notion of preservation and conservation of forests is reflected in their strong religious beliefs and ecological and economic relations with plants, trees and forests.

Conclusion

In the above description I have given a precise account of three sacred forest complexes, viz., Atharadeuli, Sapangada and Minanjhola with which many tribal religious beliefs are associated. Trips to these places are regarded as pilgrimages. Atharadeuli of Hill Kharia region and Minanjhola of the Saora region are places with which both tribal and non-tribal religious beliefs are associated. In the above descriptions three major aspects, viz., mythical, ecological and economical importance of such sacred complexes have been pinpointed. All these aspects are intertwined in the network of tribal social processes, religious beliefs and cultural patterns for which the forest and vegetation are preserved and conserved. It is evident that the mythical and religious attributes of the forest patches have designated such forests as sacred complexes. The tribal notion of preservation and conservation of biodiversity as a whole is embodied in all these attributes of the forests. The tribals are the indigenous people who have maintained a symbiotic relation with the forests around them. Religious approaches of tribal conceptions in deifying specific plant species according to their strong conviction could successfully help in serving and conserving the forest. For instance, the Hill Kharia leave patches of forest where human interference is completely prohibited. Such patches of forest are called Zaheera, meaning, a sacred grove. In the Zaheera Sal species with other flora are preserved. Kuttia Kondh and Saora do not have a notion of a sacred grove, but the forests for which they have religious attachment are better preserved in the name of gods and goddesses. Kuttia Kondh give religious importance to Sal, Asan and Banyan (*Ficus benghalensis*) species whereas the Saoras give importance to Neem trees. In the Zaheera, worship is offered to Prakriti (nature) which indirectly indicates the Kharias' love for the plant world and environment around them. In the Sapangada worship is offered to Sapangada Penu, who is supposed to be the creator of Kuttia Kondh's plant world and animals. Minanjhola is significant for Lord Jagannath was known as Maduja — the Saora god and Lord Shiva.

In all these three places the indigenous vegetation is well-preserved, perennial water streams are still in the original condition, the animals roam freely in such areas as they are not disturbed by human interference. Climax forests, where the sacred groves are particularly diverse in species of trees, are helpful for herbivorous animals who specifically depend on trees. Apart from the preservation of rare species the sacred groves may be serving the function of preservation of biological diversity even in the case of more common species of trees.[6] Forests are hardly

6. Gadgil & Vartak, 1975: 314.

much exploited when religious restrictions on certain beliefs prevail among tribals.[7]

As deforestation has been taking place at an alarming rate in many areas, such forests have come to be the only remnants of the indigenous forests in a number of cases. Because of the absence of human interference least it becomes sacrilegious, the sacred forests support the climax vegetation appropriate for their particular locality. Such a climax vegetation is very rich in species of trees, climbers and epiphytes. As such these sacred forests help in the preservation of plant species which are of great economic and ecological significance.

For preservation and conservation of biodiversity, practice of forest management in a scientific way must be integrated with the practice of traditional management of indigenous people, particularly the tribals. It appears to be a sound way of ensuring the rich floristic diversity of the forests. The tribals who are the legacy holders of the forests have appropriate practical knowledge, perceptions and observations about the ecodynamics, interaction of flora and fauna, economic aspects of the diversity, and so are quite knowledgeable about the forest. They must be included or their perceptions, feelings, ideas, values must be collected through interdisciplinary research programmes so as to plan scientific management practice of forests in a way that suits the ethos and sentiments, cultures and traditions of the tribal people.

References

Bhandari, J.S.; Subhadra Mittra Chana (1992): Forests & Tribes. A macro level prospective, *Mainstream*, Vol. 30, No.49.

Dash, J. (1992): Ecology & Techno-economic Development of the Forest Based Groups of Similipal Hills. Ph.D. Thesis, Utkal University.

Daspatnaik, P.S.(1984): Ownership Pattern, Land Survey & Settlement and its Impact on the Dongiria Kondhs of Orissa, *Adibasi*, Vol.XXIII, No.4,THRTI Bhubaneswar.

---

7. Patnaik & Jena, 1995.

Elwin, Verrier (1955): *Religion of an Indian Tribe*, OUP, Bombay.

Gadgil, M. & Vartak, V.D. (1975): Sacred Groves of India — A Plea for Continued Conservation, *Journal of Bombay Natural History Society*, Vol.72(2).

Gadgil, M. (1983): Conservation of Plant resources through biosphere reserves, In: S.K. Toria & K.L. Mehera (eds.), Dept. of Env., Govt. of India.

Haines, H.H. (1921-25): *The Botany of Bihar & Orissa*, Abelson & Co., London.

Jain, S.K. & Hajra (1979): Economic Relationship between People and Forest Flora — Man and Forest — A New Dimension in Himalaya, In: Krishna Murti Gupta & Des Bandhu (eds.): Today & Tomorrow's Printers & Publishers, New Delhi.

Mishra, H.K. (prepared): Working Plan for the Reserved Forests of Baliguda Division, 1981-82 to 2000-2001, Part I & Part II.

Patnaik, K.K. & M.K. Jena (1995): Conservation of Bio-Diversity : A Tribal Perception of Similipal Biosphere Reserve, In. R.C. Mohanty (Ed.): *Environmental Change and its Management*, Kamala Raj Publications, New Delhi.

Patnaik, N. (ed.) (1989): *The Saora*, THRTI, Bhubaneswar.

Pradhan, P. (1983): Social Forestry : *An Adivasi viewpoint, towards a New Forest Policy — people rights and environmental needs*, Indian Social Institute, Delhi.

Tiwari, D.N. (1991):Tribal Development Strategy in Sal Areas, In: *Journal of Tropical Forestry, Society of Tropical Forestry Scientists*, Jabalpur, Vol.7, November- IV.

Tylor, E.B. (1991): *Dictionary of Anthropology*, Goyl Saab, Jawahar Nagar, Delhi.

13

Forests, Tribes and Resources

Sricharan Behera

Considerations of management and conservation

PROBLEMS related to forest centre around two major issues: that of commercial use of forest produce and that of ecological preservation. These can be viewed at two levels: the macro-level involving a global distribution and use of resources and the micro- level involving the dynamics of interaction between forest dwellers and their immediate environment. In the process of civilisation two types of human population developed: those who regard the forest as not habitable and whose relationship to the forest is marginal and those who live within the forest in a more or less symbiotic relationship subsisting on the use of the forest produce. This second type of people are the tribal people who have lived amidst hills and forests since times immemorial. They are regarded as forest-dwelling people. A tribe is a social group usually with a definite area, dialect, cultural homogeneity and unifying social organisations. It may include several sub-groups such as sibs or lineages. A tribe ordinarily has a leader and may have a common ancestor as well as a patron deity. The families or small communities constituting the tribe are linked through economic, social, religious, family or blood ties (Tylor 1991).

The tribals under traditional conditions maintained a relationship with the forest in which most of the resources were used in a sustainable way. But nowadays the commercial use of forest resources by both urban and tribal people has led to large-scale deforestation reducing such resources considerably. The considerable ecological wisdom stored in the traditional cultures is hardly being used effectively for the management of the resources.

Forest and tribes are two sides of a coin. Both are significantly interrelated. Forefathers of these tribals settled within the forest and have been surviving on forests and other available natural resources in the forest. As a result their economic, social and cultural practices closely revolved round the forest life-cycle and they became the foster children of the forest. For centuries, forests have ensured a steady supply of food, fuel, fodder and other requirements of tribals. Besides, traditionally, the people had built a culture and a social system that kept a balance between human needs and the carrying capacity of the forest. The tribal people have always considered the forest a renewable resource. The tribal people in the past maintained a constructive dependence on the forest considering it as a source of many items for their material culture. No doubt a forest obviously caters to the biological, emotional and economic needs of tribal people in many ways. Their utmost dependence on forests is primarily for food. The technology and pattern of food collection from the forests and the tribal work tradition in the forest give an idea of the prevalent occupation and multifaceted relations of the tribals with the forest. The forest is important to them as a living space, as a resource base, as the 'harvest ground' of a specific socio-cultural identity. It appears as if the existence of tribal communities without the forests is impossible. Tribal communities therefore, in most cases, are described as forest-dwellers or forest-dependent communities.

Tribes in India can be divided into two broad categories depending on their habitation, namely, people who are living in the agency[1] and forest tracts on the one hand and people who

1. Agency tracts were remote areas which under British Rule came

are living in the plains of both scheduled and non-scheduled areas on the other. The tribal communities living in the vicinity of hilly terrains and forest tracts are highly religious, exposed to magic and traditional in their economic pursuits depending upon forests. But the tribes living in the plains along with non-tribal people are more acculturated in their behaviour and adjusted to the mainstream in their economic outlook due to their constant contact with the market. Tribals are found at four different stages of economic development, viz., food gathering, pastoralism, shifting cultivation and settled cultivation. Though the tribes are classified into four major groups on the basis of their predominant economic system, these categories are not exclusive. The food gathering economy mostly revolves around forest wealth. But some food gatherers here and there practise settled cultivation, while the pastoralists and shifting cultivators resort to settled cultivation. However, all these categories depend upon the forest. They have evolved a way of life which has been woven around forest ecology and forest resources for centuries.

The tribal people are emotionally attached to the forests. Their deep-rooted sentiments are in association with forests, their religion and folklore revolves around the forest. The forest is closely associated with various life-cycle activities of tribals right from birth to death. The tribals are so attached to the forests that they worship some trees in the forests as gods and goddesses. Their folklore is full of references to the forest and there are several rites and rituals which are performed under trees. Till modernisation crept into the tribal areas, a vast majority of tribal populations inhabiting the hilly and forest tracts used to survive on tubers, roots, fruits and leaves or various food varieties available in the forest, supplemented by game. As time passed on, some advanced tribal groups engaged in shifting and settled cultivation to get an assured food supply for their sustenance. Forest produce were used for supplementing the grains from

→ under the direct charge of the agency appointed by the Governor-General in Council for the suppression of human sacrifice. Gazetteer of India, 1983:42.

agriculture during the lean seasons as the returns from crops hardly sustained them for a few months in a year.

Looking through the cultural lens of the tribals, it is very much perceptible that their indigenous knowledge of maintaining a livelihood has emerged from their interrelations and interaction with the forest world. This gives an idea of the tribal people's long interaction with the forest from which their cultural identity and their understanding of a social resource derived from the existing natural resource have developed. The condition of natural resources in the world today reflects the prevailing preoccupations with technology which is based primarily on the exploitation of natural resources, of forests, lands and ocean. The forest-dwelling tribal communities, keeping themselves confined to the world of forest-experienced the maximum sustainability of their social resources.

During the past few decades after Independence the tribal areas have been exposed to outside contact with the rapid implementation of development plans that saw laying of roads connecting the inaccessible terrain and dense forests with the plain areas. Consequently, traders from the plains started migrating into the tribal areas introducing barter economy to the hitherto subsistent tribal societies. They used to supply essential commodities like salt, tobacco, kerosene and dry fish to the tribal people in exchange of agricultural and forest produce. This phenomenon generated a spirit of competition among the tribals to collect more and more forest produce to acquire adequate quantities of essential commodities. Realising the commercial value of various minor forest produce items, the traders have introduced cash in transaction with tribal people. The result is that more and more items of minor forest produce of commercial value have been identified by the tribals in the forests and are collected on a large scale. Initially exchanged on barter basis they were later sold for cash.

With the increasing demand of a market for minor and major forest produce, the tribal tradition of work changed. To cope with the market situation, often catalysed by tradesmen who trade with forest produce and act as intermediate people between the market and the forest, the tribal people brought about timely changes in their occupational patterns. This brought major changes in the tribal way of living and maintaining their livelihood. Slowly people took up agriculture, yet continued shifting cultivation and exploited forest produce as a side practice. The main occupation of a household is defined as that which bestows maximum income in relation to other occupations of the household. Tribal people are generally engaged in one or the other economic pursuits throughout the year. While larger landholding households are preoccupied with agricultural operations, landless and small households are engaged in avocations like agricultural labour and non-agricultural labour. However, almost all the households irrespective of the size of holding depend upon forests for collection of minor (MFP) and major forest produce. The collection of MFP differs depending upon the time and personnel available in the household. Quite often a household is engaged in three or four subsidiary occupations, since any single occupation cannot sustain the members of the household throughout the year. Since the tribal economy is predominantly agriculture-oriented, major income is accrued from cultivation followed by agricultural labour, collection and sale of MFP, forest labour and other non-agricultural wage labour and petty business.

An understanding of the dependence of forest-dwellers who are by and large tribals on forest resources for their economic survival and for the retention of their separate entity as an ethnic group, would help us to recast our attitude towards forest and the people in its surroundings. Socio-economic and religious practices of the forest-dwellers are closely interwoven with the life-cycle of the forest. In this paper I take the privilege of highlighting the ethnic ideas of a tribal group named Lanjia Saora in Orissa. The main occupation of the Lanjia Saora is shifting cultivation supported by paddy cultivation on terraced land and collection of minor forest produce. In central Orissa the scope of wage earning is very much limited. As a result some of them emigrate to earn their livelihood outside.

Economic dependence on tree-plant resources

Due to the absence of irrigation facilities coupled with poor water retaining capacity of soil, the Lanjia Saora grow a variety of rain-fed crops by shifting cultivation together with plain land and terrace cultivation. In the absence of a regular remunerative alternative occupation, they cannot but fall back on forest resources for their survival. Low productivity from agriculture has been partly compensated by natural rich forest. The people living deep inside the forest depend largely on forest resources like stems, tubers, fruits, leaves, flowers, herbs and animals for food, shelter and medicine. With the emergence of money economy and its penetration even into remote places, a part of their cash requirement is met from minor forest produce. During lean periods when they do not have agricultural produce at their disposal the Lanjia Saora depend upon the food available in the forests. They collect Mahua (*Madhuca longifolia*), Mango (*Mangifera indica*), Sal seed (*Shorea robusta*), Bamboo shoots (*Bambusa sps.*), Mushrooms, Jackfruits (*Artocarpus heterophyllus*), different types of wild tubers from creeper plants and other varieties of wild tubers, wild fruits and herbs. They also collect and store dried leaves, flowers and fruits to use as food in times when those are not available in the forest.

The most important of all the available trees in this area for consumption is the Sal tree. Sal leaves are used to make food plates, bowls and local cigars; its logs are used as poles for houses and for a number of agricultural implements. The seeds are used for food and oil is extracted, which is in great demand for making soups, sweets, etc., while its resin is collected and sold in the market. Sal also matters in social and religious practices. Leaves with a little rice coloured with turmeric are used as invitation letters for deities during invocation in a worship. In short, a Sal tree has a multiplicity of uses for forest-dwellers: economic, ritual and cultural. This explains why the tribals are so emotionally attached to Sal trees and are reluctant to cut them unless otherwise compelled.

Similarly, many other trees are also regarded as important resources. Let me cite some examples of the Mahua tree (*Madhuca longifolia*). Liquor from Mahua flowers is very much favoured by the forest dwellers. Irrespective of age and sex liquor is an important part of their food. Particularly Mahua liquor has a major function in their socio-cultural events. Mahua flowers are used as food and beverage. When no other food is available, Sal seeds with a few Mahua flowers make a tasty and nutritious food. Oil extracted from Mahua seeds is used for cooking and other purposes.

To obtain the local liquor (toddy), Sago palm (*Caryota urens*) and Date palm (*Phoenix sylvestris*) trees are regarded as very valuable and the tribal people always give importance to the preservation of these trees. For food, Mango (*Mangifera indica*), Jackfruit (*Artocarpus heterophyllus*), Guava (*Psidium guajava*), Kendu (*Diospyros melanoxylon*) and many other trees are held in high esteem.

The vegetation around Saora villages, comprise Tamarind (*Tamarindus indica*) trees, followed by Mango (*Mangifera indica*) and Jackfruit (*Artocarpus heterophyllus*) trees. No doubt, Saora settlements are made on cleared patches of forests where Sal is the dominant indigenous vegetation. But at present Tamarind, Mango, Jackfruit, Mahua, Karanj (*Pongamia pinnata*) and Jamun (*Syzigium cuminii*) trees indicate that the Saoras made a selection of species which they prefer in their surroundings. Tamarind is a very important addition to their regular food. Mango and Jackfruits are major crops on which the people subsist during lean periods. Mahua and Karanj are oil seeds. Karanj is an important tree for rituals. The belief goes that if a dead body is not cremated with Karanj wood, then its life may return again to torture the living kin. The Karanj wood is also important from the point of view of the ornaments and decorative parts of the Saora material culture.

Tree and plant resources are given highest importance as renewable resources. When a family owns fruit and other trees, women and children can collect MFP together since the trees are close to their houses. When a family owns no trees, then it depends only on lands. More powerful families monopolise the trees that are close to the village and the poorer sections have to go far away into the forests. For the Lanjia Saora the utility of their food trees as a resource, be they near the village or in the forest, is always appreciated. Therefore, the people, as my observation and interviews with people revealed are not much interested in monocultural plantations and big horticultural gardens.

Fruit bearing trees like Mango, Jackfruit, Tamarind, etc., are numerous in the forest near Lanjia Saora settlements. Therefore, they do not like to undertake plantations of such species supplied by development agencies. On the other hand they have small patches of land where they cultivate mixed crops (in swiddens) and grow paddy in paddy lands. Development agencies are interested to promote horticulture in swidden plots (under cultivation or abandoned). People cultivate their subsistence crops in swiddens and they hardly have more patches of lands to spare especially for horticulture. There are two kinds of horticultural activities. One is plantation of various species of fruit bearing trees in homestead lands and the other is horticulture of cashew nut (*Anacardium occidentale*) in abandoned or less productive swidden plots. Economically-minded Lanjia Saoras have more paddy land, take interest in paddy and spare their swidden plots for cashew nut plantations. But those who are marginalised do not like to undertake cashew nut plantations lest they lose their subsistence crops. In the debate whether subsistence crop or cash crop is important, two different economic classes of people emerge. Most who are marginalised thus do not have good interest in monotype or diverse type of horticultural plantations. Again, no fruit except the cashew nut fetches a good price in the market compared to the labour input.

Soara concepts of social and natural resources

The Lanjia Saora people take the forest as a resource. I opt to give here their ideas of a renewable resource, in a rather simple way with reference to the peoples' terms. Resource is any part of man's physical environment, living or non-living or any human skill and knowledge that can be utilised in the production of goods and services. Man, since ages, is deriving all his needs from natural resources, viz., earth, water, forest, lands, etc. The principal necessities of man are food, clothes and shelter. The resources used to provide these three prime needs are always given importance since times immemorial. Since human beings evolved, they started depending upon the forests as the major source of most of their household needs and developed close relations to the forest. The resources that are used for their household needs, minor economy, food and shelter which I interpret as 'social resources' are available in forests and represent 'free goods' at the disposal of those who make use of it. With the growth of population and advent of civilisation the demands and pressure on resources increased so that the exploitation of forest resources became a problem in some areas. Now, civilisation has reached a point where human wants and demands at the national level are unlimited and resources are limited.

The Saora understand that making use of resources in a socially appropriate way provides them with 'social resources' which are part of the natural resources around them. To describe their understanding of a social resource I must take reference to certain Saora terms which are presented in Table 1. 'Janangan' shows their understanding of land as a resource base or a source of many resources. 'Janangan' conceptually refers to land resources. Baroon, Saroba, Baseng, Aagari, Kutta which mean swidden plots, paddy lands, foot hill lands, plateaux and homestead lands once have been parts of forests. Janangan is the most important 'social resource' in the perception of the Saora. This shows their dependence on forests or cleared patches of forests for agriculture and food provisions. Baroon is the place which refers to the swidden plots in the middle of a hill where the people practice shifting cultivation. In the Aagari (places completely barren or plateaus) they cultivate specific crops like Sorghum, Bajra (*Sorghum vulgare*) and Little Millets. In the Saroba

TABLE 1: Classification of Social Resources

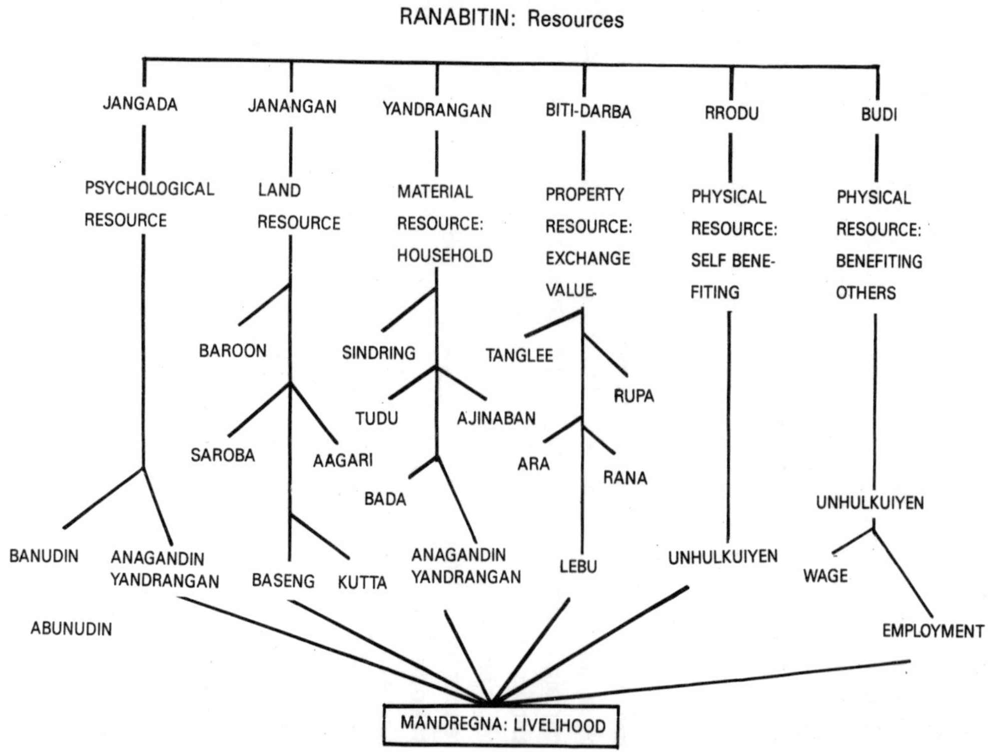

Note

Aagari: Plateau
Abunudin: Community knowledge
Ajinaban: Usable things
Anagandin yandrangan: Skill
Ara: Trees
Bada: Gun
Banudin: Individual knowledge
Baroon: Swidden plot
Baseng: Foot hill land
Biti-darba: Property
Budi: Man power/utilisation by others
Janaban yandraam: Tools and implements
Janangan: Collective landed property, land resources
Jangada: Collective knowledge/wisdom
Kutta: Homestead land
Lebu: Money
Rana: Gold
Rrodu: Man power: self-utilisation
Rupa: Silver
Sindring: Dress materials
Saroba: Paddy land
Tanglee: Cattle
Tudu: Bamboo makes
Unhulkuiyen: Measurement of labour
Yandrangan: Material resource/Household

which here refers to plain paddy lands, they crop seasonal paddy; one Kharif (summer crop) and one Rabi (winter crop). In the foothill lands they cultivate cash crops like Til (*Guizotia abyssinica*), mustard, Niger (*Sesamum orientale*). In the kitchen garden vegetables are grown. The trees they own personally are found at different locations. But the trees are considered in another category of their social resources. Biti-darba literally means property and includes materials like gold, silver, money, etc., which have an exchange value. Cattle and trees are put in this category. Here the Saora give trees and cattle value in terms of currency. They mean that cattle can be exchanged for cash and similarly trees or produce of the trees can be sold. Cattle have regenerative power and so have the trees. These are perceived as renewable resource because they give regular returns in terms of cash and food. Here only such trees are taken into consideration that are personally owned. Trees in forests are not personally owned, therefore these are not considered here.

Table 2 gives an understanding of the mobility of Biti-darba, i.e., property and materials. Here Biti-darba is mainly classified into two broad categories: Damnien Bitin (personal property) and Arbatinjii Bitin (collective property). Property, whether individual or collective, is further classified into five categories:

1. Barra-ra-bitin — Mobile resource
2. Er-barra-ra-bitin — Immobile resource
3. Abnar-ganab-dil/ Abnar-sanab-ja — Conversion of one thing into another
4. Abna-ar-bitin — Renewable resource
5. Barra-ra-adena-bitin — Unusable resource

From all these properties (personal or common) ideas on Anabniana, i.e., management and Abnyanatai, i.e., maintenance can be derived. Through Denisam (evaluation/assessment) the Muchchhale (loss) and Lavom (profit) can be accounted. This

Table 2: Classification of Social Resources

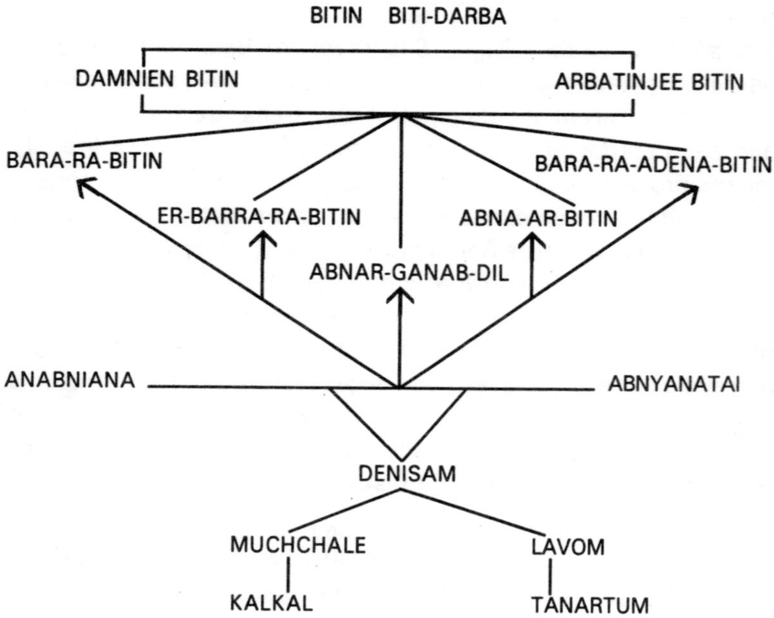

Note

Abna-ar-bitin	:	Renewable resource
Abnar-ganab-dil / Abnar-sanab-ja	:	Conversion of one thing into another
Abnyanatai	:	Maintenance
Anabniana	:	Management
Arbatinjee Bitin	:	Collective property; not household property
Barra-ra-adena-bitin	:	Unusable resource
Barra-ra-bitin	:	Mobile resource
Bitin, Biti-darba	:	Property
Damnien Bitin	:	Personal property
Denisam	:	Estimation/Calculation/Evaluation/Value analysis
Er-barra-ra-bitin	:	Immobile
Kalkal	:	Crisis/Deficit
Lavom	:	Profit
Muchchhale	:	Loss
Tanartum	:	Surplus/Buffer/Conservation

leads either to a Lalkal (famine) or Tanartum, i.e., a buffer or surplus stock. Tanartum is always assured when a person owns good tree species. Trees in the forest and trees personally owned are always regarded as a Barra-ra-bitin, i.e., mobile resource as well as an Abna-ar-bitin, i.e., renewable resource.

In Table 3, there are three important resources: Lebu (money), Bunudin (knowledge) and Rrodu (manpower) which collectively determine the making of Biti-darba. These are the chief inputs of management and maintenance practices. Biti-darba is classified here in terms of Eraing Bitin, i.e., renewable or inexhaustible resources and Si-Eet-Bitin, i.e., perishable or exhaustible resources. The Saora consider Soroba (paddy lands), Lobo (land/earth), Gada or Tulab (forests), Jada (stream/water sources), and Ara (trees owned personally or commonly as renewable resources). People consider from Saroba, agricultural produce from Lobo, useful vegetation, materials and animals from Gada or Tulab, drinking water and water for irrigation from Jada, food, shelter, cash from Ara as perishable resources. However, their resources are mobilised through their indigenous system to get the outputs that are enlisted in the categories of exhaustible or perishable resources. Through better management, the sustainability of a resource may be preserved. The Saora believe that through appropriate utilisation and better management a perishable resource can be converted into a renewable resource and on the other hand misutilisation and mismanagement gives the reverse effect, i.e.; a renewable resource becomes a perishable resource. Some Saora perceive their crisis as an outcome of misutilisation and mismanagement of renewable resources.

Deforestation and crisis in resource use

The Saora are reluctant to admit that they are solely responsible for the deforestation. They accuse the non-tribals whose material demands from forests are higher compared to those of the tribals and do not easily admit that deforestation is caused by shifting cultivation only. The tribals under traditional conditions maintained a relationship with the forests in which resource use was

TABLE 3: Classification of Social Resources

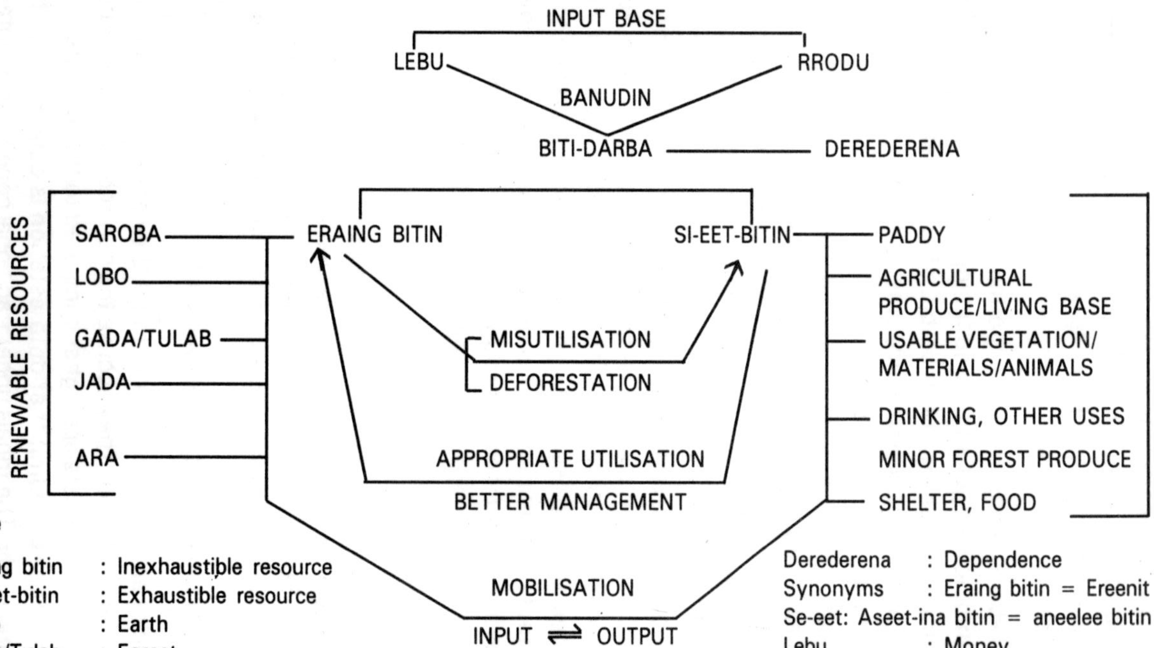

Note

Eraing bitin	: Inexhaustible resource
Si-eet-bitin	: Exhaustible resource
Lobo	: Earth
Gada/Tulab	: Forest
Jada	: Water resource
Ara	: Trees
Saroba	: Paddy land

Derederena : Dependence
Synonyms : Eraing bitin = Ereenit bitin
Se-eet: Aseet-ina bitin = aneelee bitin
Lebu : Money
Rrodu : Manpower
Banudin : Individual Knowledge
Biti-darba : Property

regenerative. Neither foraging nor shifting cultivation had any large-scale irreversible degenerative effect on the environment. The popularly accepted view was that the tribals should be left to continue their traditional mode of life in the forest, although it is true that commercial use of forest resources has led to large scale deforestation in recent times. Conditions of modernity, a higher rate of demographic growth and increased consumption goals of the people have made the traditional methods of foraging and horticulture uneconomic options even for the population traditionally engaged in them.

The ideas of ecological preservation, particularly the preservation of natural resources, have given rise to completely new ideas regarding forests and their native dwellers. People who were earlier regarded as causing the destruction of forest resources may now become their conservators.

References

Bhandari, J.S. & Subhadra Mittra Channa (1992): Forests & Tribes: A macro level prospective, *Mainstream*, Vol. 30, No. 49.

Gazetteer of India (1983): Orissa (Boudh-Khondmals). Department of Revenue, Government of Orissa.

Tylor, E.B. (1991): *Dictionary of Anthropology*, Goyl Saab, Jawahar Nagar, Reprint, Delhi.

14

Symbolic Representation of Being in the Sacred Art of the Saora of Orissa

Padmini Pathi

The Saora myth of coming into being

THE Saora are a tribal community concentrated in the southern part of Orissa. They are also known by many other names like Sabar, Savar, Sora, etc. This paper deals specifically with the Lanjia Saora, the primitive section, and the Shudha Saora, the Hinduised section of the tribe, with reference to the structural, functional and symbolic aspects of their religion.

The Saora trace their origin from 'Kureitung Katabir', their cosmogonical myth preserved in their oral traditions. The cosmogonical myth of the Shudha Saora and Lanjia Saora varies to a great extent. The Shudha Saora hold that Kureitung, from where they evolved, was a hollow wooden trunk. But for the Lanjia Saora Kureitung was a hollow gourd from which they evolved. The myth of the Lanjia Saora and Shudha Saora coincide at a particular point where both of them believe that they came into being from Kureitung. Kureitung Katabir is the only reference from which the Saora know about their coming into being as social beings.

A comprehensive account of the cosmogonical myth narrates that in the beginning of the universe there was nobody else other than a man and a woman. The first person regarded as the sole ancestral person is known as Kittung. His counterpart female is known as Radab. They were living inside Kureitung. Through an illusion made by a Supreme Being (Korangulu — as in the Shudha Saora myth of origin) they started their conjugal life. Through their act of sexual procreation further people came into being. Kittung and Radab are thus the first known couple and the first and primeval entities from whom the Shudha Saora people came into being. They therefore consider Kittung and Radab the first Saora people.

One reason may be given as to why Kittung and Radab are regarded as Saora. From this the etymological meaning of Saora is derived. As the people explain, the compound of the word Saora is the synthesis of two terms 'so' and 'ara'. The former means 'hidden' and the latter means 'wood'. The first couple (Kittung and Radab) were hidden in a hollow wooden trunk. Further Kittung and Radab, the first known human beings, discovered by the Supreme Beings (Barangdeu, Korangulu and Sisitamroga) from Kureitung, justify the etymological meaning of Saroa. Kittung and Radab in later times became social beings on the earth and became known as Kinjinangbui and Kinjinangboi. Both Kinjinanbui and Kinjinangboi in fact are rather specialised social beings compared to the average Saora folks for their shamanism. They are empowered by the female Supreme Beings (Allamai, Sintemai and Gangamai Raudi) that give them a secret power of communicating with and coordinating different planes of the living space of different beings. Finally after their death they become spirits living in a separate living plane.

In the above myth about the origin of the Shudha Saora, four stages of the Saora after coming into being are well-marked. The first stage marks the discovery of the human beings living in Kureitung regarded as Saora later on. The second stage marks their conjugal life stage and the beginning of hereditary rights. The third phase is the beginning of shamanism for the community's welfare: to preserve the community's well-being through the act of religious performances. The fourth phase is the stage when the physical form is lost, the visible entity becomes invisible (after death the dead body is burnt) and the transfer of the soul from a human being to a spiritual being takes place. The Saora thus trace their origin and the evolution of mankind, the growth and development of race and the state of life and death from their myth of origin in which their world-view is emphasised. Kittung is therefore the first person from whom the Saora people are descended. One day Kittung became very emotional. In his mind it was settled that once he died, he would have to leave his descendants. With this thought the consciousness of the mortality of beings came to the Saora mind.

In the Saora myth of origin Kittung is positioned at the peak of the pedigree of human evolution. For long the Saora took man as immortal. This idea was based upon the fact that Kittung lived for long on earth. The Saora could not believe that life was mortal. They had and also to some extent have the self-conviction that men in primordial times lived for ever and did not die. They understood the animal language and lived in peace with all animals. They had not yet to work, had plenty of food and its effortless acquisition guaranteed a carefree life to them. They had no idea or knowledge of sex and procreation. One day through some request they put themselves into opposition to god by transgressing a commandment, by curiosity, or similar acts. Disorder followed. The paradisiac condition came to an end. Man began to be troubled by worries about his food. Through a first murder the original immortality of man was taken away. Their consciousness of the mortal nature of man came through the death of Kittung by an accident, when a Saora killed Kittung with an arrow while the latter was enjoying a swing on a Siali (*Bauhinia vahlii*) creeper. The arrow hit Kittung and he died. This was the first event when the Saora experienced and became aware that man was mortal. The story does not end here. After the death of Kittung his soul left the physical body. It was reborn in an incarnation, after the Saora killed Kittung by his arrow. The divine nature of Kittung after his incarnation was called *Madusa*. The Saora perceive this second stage of life, where a transfer and transformation of the soul takes place, as the spiritual stage of being.

The Saora thought on life, death, mortality and transfer of the soul from one body to another shapes their understanding of existence of the soul in different entities. They believe in the the existence of the soul in two forms: living and non-living, the living referring to the earthly living beings and the non-living referring to spiritual beings. For these two stages of existence of the soul, two terms have been coined by them: Saora and Sonum.

The Shudha Saora's notion of coming into being shows a two-step evolution. In the first step Kittung and Radab are known as the only living beings. Since it is not mentioned anywhere in the myth how Kittung and Radab came into being, we may assume that they have not taken birth from anything but rather they are presenting 'Being' itself. The second step of Saora evolution is based on the relation of sexuality and procreation: Kittung and Radab stand as the ancestral masculine and feminine entities from whose act of procreation the Saora people came into being. The Saora develop a conviction from this: they are mortal because they descended from Kittung and Radab who in fact reproduced the Saora people through biological means.

In the Lanjia Saora cosmogonical myth, a huge population of Lanjia Saora was born from the couple in Kureitung. Everybody remained confined in the gourd. A devastating rain came. The whole earth was submerged. Then the earth dried. The vegetation grew. An eagle and a bear are mentioned who took great pain in bringing about vegetation on earth. The first plant that grew on the earth was *purpuri* – a grass botanically known as *Cynodon dactylon*. Then the people came out from Kureitung and lived on the earth. From then onwards the people became social beings.

The comparative view on the myths of the Shudha Saora and Lanjia Saora illustrates that in either case the first beings (human beings) were derived from Kureitung. From them the present-day Saora evolved. The question, however, remains unanswered: who created the first people living in Kureitung? Both the Lanjia Saora and Shudha Saora claim that it was Sonum. In my terminology I put Sonum as the Supreme Being. In their cosmological (Saora Jagat) perception the Saora hold that the Supreme Being came into being from an undefined larger phenomenon. Sonum literally refers to 'spirit'. So Sonum is not specific here, at least not specified by a name. Hence, I interpret that Sonum or the Supreme Being in the understanding of the Saora is an invisible entity. Further the larger phenomenon — Being — is not clearly defined.

Every society is in some way or the other either masculine or feminine in its orientation or more exactly, the culture-ideal is determined either by man or by woman, which in extreme cases leads to the fact that either the male or the female type appears as the culture-ideal. The Saora society shows masculine orientation because the Saora take reference of their ancestral man Kittung. Both Lanjia and Shudha Saora are descended in a patrilineal line because in latter cases they take the name of the man of the sole ancestor couple. On a still more abstract level the Saora perceive the entire world and all of its beings as created by a 'Great Creator' and all the entity of the world as the immutable unfolding of an ultimate pattern of destiny. The Saora perceive this 'Great Creator', in my terminology Being, as an invisible supernatural (Sonum).

The Lanjia Saora sometimes take the name Kittung to specify their originator, Sonum. That mankind descended from Kittung is a conviction held in a legendary tale of the Lanjia Saora. In the village Dungdungar I heard the following tale from a Lanjia Saora informant: Once Kittung painted an icon on a inner wall of his house. He painted a man and a woman in white colour, with a desire to bring to earth the human beings. Ignorant of how human beings can be created, Kittung asked Uyungsum (the sun god) to help him. Uyungsum instructed Kittung to cover the drawing of the man and woman with Siali leaves. He also directed Kittung to offer blood from his body to the man and woman represented in the icon. Kittung offered blood to the pictures by cutting his hand on the seventh day after the drawing was made. He smeared the blood on the picture of the man and woman. Some days after the offering of the blood a girl emerged from the wall painting and on the next day a boy emerged out of the drawing. When the boy and the girl reached their marriageable age, Uyungsum organised their marriage and from them the other people came.

This legend is also reported by Elwin in his monumental work *The Religion of an Indian Tribe* (1955: 402). From all these legends and myths cited, it is clear that the sole ancestral couple emanated from the will of the 'Greatest Creator' or Being! From this ancestral couple came the knowledge of sexuality and procreation.

The 'Greatest Creator' or in the widest sense 'Being' is known to the Saora by the name Kittung. Kittung is their creator and originator. The Saora people's evolution has taken place by biological means. But Kittung's evolution is not biological. Kittung's birth is mythical and mystical, spontaneous and abiological. From the Kittung's mysterious coming into being, the Saora conceive and perceive that there is yet another greater potential in the cosmos which is the originator of the first life. This greater potential is the sun which is regarded as the supreme power, the experience of the conception of life. This is their view of Being. Before Kittung took birth the entire universe was created by some other supernatural force. Hence the Saora in their conception of the greater Being view Being as a broader whole that contains the cosmological components like earth, sky, moon, stars, water, land, fire, hills, mountains, stones, forests, etc. The cosmological components are always fixed to their respective positions. But the living beings are not fixed as they are. Simply, when the cosmological components are not changing or dying, it makes sense that such things are immortal and the living beings for whom birth and death is the truth, are mortal. Based on this, the Saora view the cosmos or the greater Being as most powerful from which Kittung took birth and later on the Saora came into being. Here the Saora divide the universal components into two sets. The first set where the greater Being is positioned and the second set where the living beings are taken into account.

Hence in the Saora cosmological perception there came out three sets of beings: the Supreme Being, the powerful celestial beings and the living beings. The Supreme Being is an invisible entity – celestial beings can be imagined and living beings and Kittung and Radab represent the first living beings. The Shudha Saora hold that the animal took birth from human urges and instincts. But the Lanjia Saora believe that animals had already taken birth before they came into being. But in either case people give significant value to Being whom they believe to be the creator of the plants, animals and human beings.

The Saora conception of anthropomorphism ultimately became limited to the understanding of Kittung, the Saora themselves and the spirits. The Saora perceive that every object or living thing has a spirit that steers or animates its behaviour; when a person dies or a house collapses, spirit and physical form have become separated. Important to the daily life of the Saora are different categories of spirits whom I regard here as free spirits. First, there are ancestral spirits watching over every member (kin) of the Saora, rewarding them when kinship norms are observed and punishing them when they are violated. But man is on the same footing as all other orders of phenomena – the birds, the beasts, plants, and insects, the mythical beings, the ancestors, all these and many more, each with its order function. The Saora believe that Kittung, the great ancestral spirit, is a person, aloof, remote and unconscious, all who powerful, who never exercises his power directly. The Saora perceive the objective world outside as a world of Sonums of different kinds innumerable, some though – not all – preserved in memory. The general conviction is that the Sonums are like the Saora people. The difference is that Sonums can exist in a number of planes. But the Saora are confined to one plane only. The Saora people believe that Sonums in their worlds live like the Saora.

It is not easy to interpret the Saora understanding of a Sonum in the exact way in which they understand it. In Saora understanding, both the Supreme Being and the dead ancestors are regarded as Sonum. But they are quite certain that Sonums have a soul and the Sonum takes rebirth or manifestations. Therefore, to avoid all confusion and debates in the identification and classification of Sonums, I explaining describe them here as Sonum beings. With this I continue describing the different status and states of existence of the being of beings.

Accordingly, different categories of beings, man-, plant- and animal-spirits, are located in different features of the landscape. As a person moves around this landscape, he may encounter Sonum beings and become involved with them. His encounters with the Sonum beings may cause illness. An illness in a living person is a reflection of a mood or attitude of a dead person, and the rite which heals a sick person is also a rite to acknowledge the claims of the Sonum beings who have attacked him.

Vitebsky, who did ethnographic fieldwork among the Saora, perceives them in the following way:

> The Sonums are enigmatic not only because they are inaccessible. After death the consciousness becomes a powerful but contradictory causal principle, nourishing its living kin through the soul force it puts into their growing crops but at the same time precipitating illness and death among them. In one aspect, the dead give to the living their continued sustenance and their very existence, yet in another they diminish their being as they impinge on them and consume them.
>
> — 1993: 9f

Considering the multifaceted actions of the Sonums, Vitebsky has classified the Sonums into two main categories: Experience spirits and Ancestral spirits. The Experience Sonum is a member of several collectives. These contain many people who have died in a similar way. In this aspect the personality and lineage roles are submerged with an emphasis on the experience of pain which one has suffered and on the aggressiveness which one tries to perpetuate. The Ancestor Sonum retains his role within the lineage and something of his personality and nurtures his descendants (*ibid.*).

In either of the above-mentioned Sonums one thing is well marked: Sonum is the spirit of the dead. The type of death determines the spirit's living space. In case of a natural death, the spirit lives in the underground and is regarded as an ancestral spirit. When the death is unnatural, immature or accidental, then the Sonum of the person becomes an Experience spirit in which the experience of his death is well-defined. It makes sense that the anima of a living being is preserved in another world in the name of a Sonum and the type of death does not matter here. A person may die naturally or in an accident. Sickness, accident, etc., are the experiences which a person gets before death. This experience of death determines the dead spirit's living space – underground or space above ground. In case of a natural death the spirit lives underground and is regarded as an ancestral spirit. When the death is unnatural, immature or accidental with an experience of type of death, then the spirit of the person (experience spirit) lives in the space above the ground. This space above the ground is the experience space where experience spirits live. For different kinds of death experiences, different layers of living space exist in the experience space. This makes me that the Saora trace the being of Sonums and the Sonum beings from their memory of the death of the person. Whether dead by experience or dead by nature, the soul of life is not lost. It is simply transferred from the human being to Sonum being. In this context of presentation of anima, it can be said that life is never lost, but simply transformed from man to spirit, analogue to the law of conservation of energy: neither created nor destroyed. Hence experience of death matters to trace the living space of the spirit and the experience of death does not matter for the conviction that a man died and became a spirit.

Categorisation of beings

In the anthropomorphic descriptions given by the Saora people of the aforementioned different types of beings, five different sets of beings, both biological and abiological, appear to be living in the entire cosmos. They are given different names like Saora, Kulba, Idai, Ilda and supernatural entities simply known as Sonums or Sonums with a prefix indicating their type as, for instance, Uyungsum. These again may be classified into different categories. Kulba, Idai and Ilda are ancestral spirits, the undefined Sonums are the experience spirits, and the Saora people are the biological beings. The Saora, plants and animals are confined to the Lobodesa, i.e., earth. The ancestral spirits who live in the underworld occasionally visit the earth, may live for a short period here but never visit the sky.

The ancestral beings and supernatural beings occasionally visit the Saoras either to reward them for observance of religious rites and rituals or to punish those through a disease who violate religious principles. Their visit to the earth determines the Saora's well-being and good life. To preserve their well-being, the Saora invite the ancestral and supernatural beings in their socio-cultural events and extend due honour to the spirits through worship. Thus the biological beings, experience beings and ancestral beings influence the Saora, plants, like agricultural crops and domestic animals in many ways.

The Saora demarcate five dimensions in the universe where they place different forms of beings. In the first place the sky is the habitat of supernatural beings, like sun, moon and many others. The space in-between the earth and sky is occupied by Kulba beings, collectively called Kulbanji who are ancestral beings with many unsatisfied, unfulfilled material desires. They live collectively and individually in different types of landscapes like rocks, hillocks, trees, streams, etc. Immediately after death, the *feuradan* (soul) of a person leaves the body and lives somewhere in the immediate surrounding. It lives like a nomadic spirit till the second mortuary rite (*Goar*) is performed. After the observance of *Goar* the soul enters into the underworld, but it still lives like a spirit without a home of its own. It shares his living space with other dwellers in the underworld. It does not have the privilege like the inmate dwellers in the underworld to have a permanent place of its own. Of course it is regarded as one among others in the underworld. Its living standard after *Goar* and before the final mortuary rite (*Karja*) is compared to the life of a slave or a servant who lives with the master and acts for the master. The soul at this stage has a lot of materialistic desires. Particularly, it thinks much about the crops it cropped during its lifetime. In its mind there always remains a greed for these crops. It desires to fetch the harvest that would cater to its need for food. The soul at this stage remains dynamic. After the third mortuary rite (*Karja*) is held for the dead, his life in the underworld becomes static. He gets a permanent dwelling place in the underground and lives with the other ancestral spirits. Hence Kulbanji are characterised as a group of spirits having unfulfilled materialistic desires, impermanent dwelling places and souls which move here and there.

The mortuary rites are important occasions for the departed soul. By their observances the soul turns to be an ancestral being. Elwin has rightly pointed out:

> (...) the Goar is the great turning point in the life of the dead. Its effect may sometimes have to be driven home as it were, by the Karja ceremonies, but for all practical purposes ..., the Goar changes the shade into an ancestor and admits him to the underworld.
>
> — 1955: 67

It is most probable, at least as I feel, Kulba and Elwin's 'shade' refer to an invisible ancestral being who by behaviour is intermediate between the Saora beings and their ancestral beings.

Table 1: Categorisation of beings

Supreme, Living, Ancestral — being

Sonum:	(Experience Spirit) Celestial: Sun, Moon Environmental: Earth Sonum, Stream Sonum, Wind Sonum, etc.
Kulbanj:	Ancestral Spirits with unfulfilled materialistic desires. People who died in accidents and through mishaps (when death is not natural, rather unnatural and premature) become Kulba (Kulbanji). Live in trees, rocks, forests hills etc. Those spirits for whom Goar and Karja rites are not performed become Kulbanji. Malevolent mostly because of unfulfilled desires.
Saora:	Living Soara people and their living inmates (plants, animals and human beings).
Idai:	Ancestral Spirits in the underworld. People died normally, Karja and Goar performed in the appropriate way.
Rauda:	Great grand ancestral spirits, some believe them to be dead Shamans' spirits. Ildas are accounted here who are supposed to be children of Shamans and their underworld dead partners. Ildas are not strictly confined to this space.

Those *Kulba* beings or shades always move here and there and try to be with Saora people, and therefore wander within the above world to satisfy their unfulfilled materialistic desires. They visit the earth to cause harm to people driven by an intention of envy and selfish motives. Whenever a *Kulba* being comes across a happy person, it becomes envious and causes some problems to the person. *Kulba* beings become successful in causing a disease or putting a bad action on a person. But it is made powerless by the shaman's act.

The Saora people have a strong conviction that *Kulbanji* are more or less similar to the Saora people in appearance. Normally they believe that a *Kulba* has a similar appearance as the Saora person who died and became a *Kulba*. Yet there is a second kind of *Kulba* beings. Those are the spirits for whom none of the mortuary rites were performed. They live permanently in the living space of the Saora. Those permanently living *Kulba* beings have relations with the dangerous evil spirit *Ra'tudsum* who is extremely malevolent.

The Saora people share their living space with plants, animals, hills and forests, which are their immediate inmates. The Saora depend upon their labour to earn their livelihood through the processes of agriculture, food gathering and hunting, and depend upon the ancestral and supernatural beings, the good will of whom yields a lot for the Saora. The supernatural beings, ancestral beings and Kulba beings casually come there. This space is considered to be their common living beings space.

The underworld immediately under the Saora's living space is the dwelling place of the ancestral spirits see Table 2. They are called *Idai* and are permanent dwellers of the underworld (*Kinarai desa*). These Idai beings maintain a social life like the Saora people on earth. They perform marriage, cultivation practices – all like living Saora people. The stability of the spirits in the Kinarai desa is ensured by the performance of Goar and Karja by their living kin for them. Those spirits have relatively few materialistic desires as compared to the Kulbanji. Sometimes they migrate to the Saora's living space, possess some person and demand something from the people. Whenever they visit the Saora people they pass through the rice pounding hole of the Saora's house. This rice pounding hole links *Lobodesa* (earth) with *Kinarai desa*. Through this hole the Saora people communicate with their ancestors. The journey of the Saora people to the underworld is made only once, when they die and with no hopes of returning to their bodies. The earth and the underworld are linked by a huge tree which the Saora shaman must clamber. Saora people, as they know from their shamans, say that the path includes dizzying precipices on the way down. In order to make this journey, sometimes the shaman's soul becomes *gadille*, a monkey.

In some of the Saora songs, an imagination of the underground is given. The path is narrow and rather metaphorical in Saora terms, literally meaning 'tight rope path', 'the impossible imbalancing path' etc. Walking to the underworld is said to be metaphorically like 'your monkey's four footed walk', 'your ape's four footed path' (Vitebsky 1993:19). The path is terrifying, but the ancestors are kind and reassuring is the Saora's view of the ancestral beings. One living space is not strictly confined to any dimension. Here the living space is the whole area beneath the sky. There the Ilda and Rauda live. Ilda are omnipresent and can live anywhere except in the living space of supernatural beings. Ilda are a special type of beings who are neither ancestral beings nor supernatural beings. The Ilda are supposed to be the children of Rauda and shamans, the former and the latter being of opposite sex. Rauda are the great grand ancestral spirits. They are given the highest position among the ancestral beings. The Saora sometimes regard Rauda as Kittung, the originator of the Saora people. The judiciary discipline is managed by the Ilda. The Ilda control crime and keep a balance of virtues and vices. Having taken birth from the sexual union between a living and a dead, they are free to communicate between the visible and the invisible world. I conclude that besides the living space there are four distinctly marked living spaces (see Table 2: a,b,d,e). With their understanding of various forms and origin of different types of beings, the Saora perceive a spatial distribution of the physical environment around. In this understanding their idea of migration and manifestation of the soul or anima of beings and the various states of anima is embedded.

Beings in Saora religion

One very important and remarkable religious performance of the Saora is the icon art, much known as the folk art of the Saora. The paintings are done on various occasions for various reasons that describe the type of contact of the Saora with ancestral and supernatural beings. Icons depict a picture of inter-relation among living beings, non-living beings, materials and objects. The man-plant-animal spirit interplay is clearly sketched in Saora religion. In my understanding it is an interplay of beings. I comprehend the beings as belonging to two categories: living

Table 2: Saora living spaces

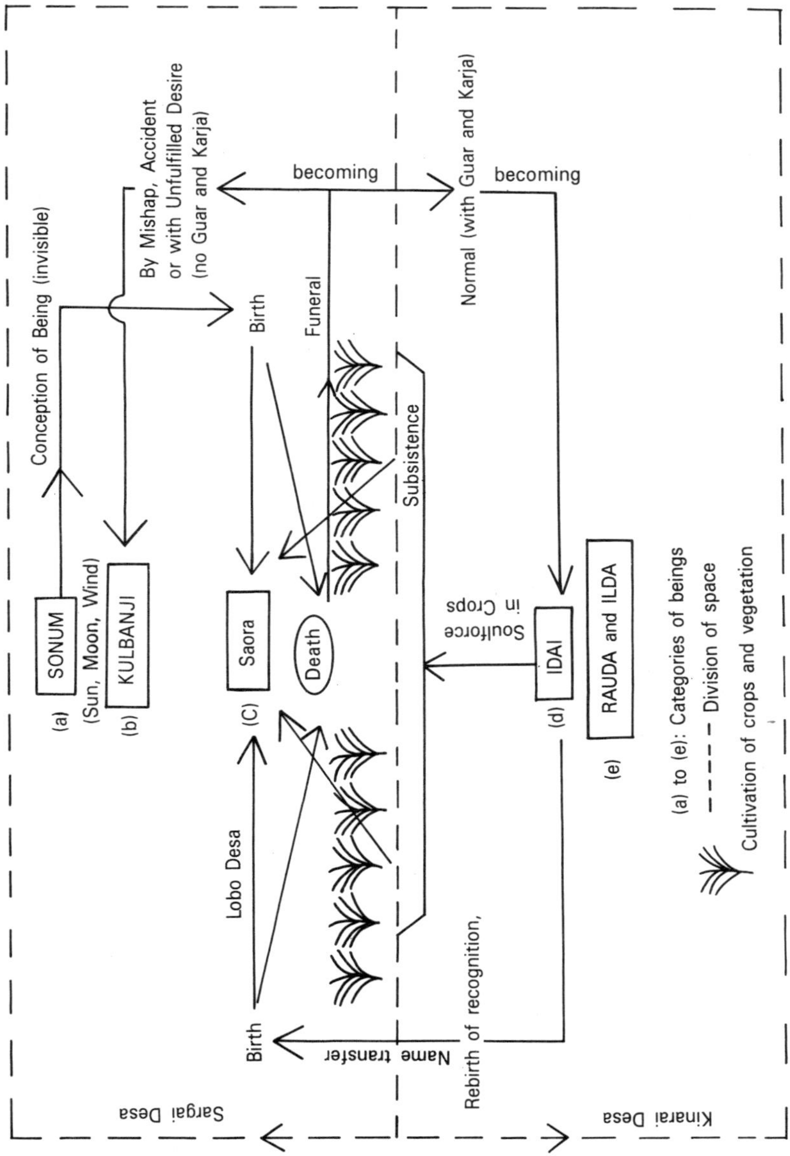

beings and spiritual beings (ancestral and supernatural). The Saora perceive their beings and understand their actions through religious performances and beliefs.

Different designs of the icons and different types of arrangements of beings in the icon provoke a social discourse of the art. The Saora fear complex towards ancestral and supernatural beings is further expressed in icon drawing. As it has always been since times immemorial, the Saora establish their good relation, usually a fellowship, with the ancestral and supernatural beings through icons. To a major extent it preserves the Saora notion of well-being, property, health, etc. In the icon, the Saora understanding of their existence, culture and the philosophical attributes of man-spirit interplay is represented. The Saora, as it seems, have an elaborate system of appeasement to protect themselves against supernaturally inflicted diseases, and an important ailment in their security measures is the making of icons, which has, however, to be combined with animal sacrifices.

Further, many drawings are associated with the dead, and it is possible to gain from them some Saora notion of their conception of life in the other world. It is not easy to understand all the details of these pictures without an exhaustive discussion of Saora eschatology, but it may be said briefly that the Saora believe that the dead go to the underworld, which is ruled by beings which can be more properly described as spirit-helpers than gods. The dead pay frequent visits to demand attention and sustenance from the living and, particularly before their funerary rites have been completed, can be very troublesome indeed. The spirit-helpers, too, often visit men and there is an elaborate system of matrimonial relationship between this world and the other (Elwin 1954: 198f.).

Icon drawings are combined with animal sacrifices as in most other rituals and performances to cure fatal diseases. "In Saora theory almost all disease is the result of interference from the unseen world...." (*ibid*.: 192). Icons are also painted to avoid or mitigate the interference of the entities from the unseen world.

Here sacrifice has a special meaning. Through a sacrifice, the life of a Saora is saved. Sacrifices are given to appease those beings whom Elwin regards as spirit-helpers. Further sacrificing an animal does not only mean offering blood of the sacrificial animal to the spirit-helper who is an ancestral being. From a talk with Prof. Vitebsky I got some idea about the Saora belief of animal sacrifice. Vitebsky holds, on the basis of the Saora belief, that an animal sacrifice means offering an animal soul for the release of the soul of a Saora which is inflicted by a disease caused by an ancestral or supernatural being. So sacrifice means to exchange a soul for another one.

The understanding of the sacrifice of an animal has links with icon art. The art area is regarded as a seat offered to ancestral and supernatural spirits in the Saora house. Here I would like to discuss a little about how the icon gathers all types of beings at one place, in a small drawing, in a dark corner of the Saora house. Through the anthropomorphic figures of ancestral and supernatural beings and through the pictures of objects and subjects, the Saora conceptualise their concept of mobility, stability and activity of beings. As I have discussed earlier, of the different kinds of beings, all or most of such beings are to be found in the icon. The anthropomorphism and symbolism of the beings are illustrated in the icon. In the context of art, I would like to mention the figurative representation of beings in the icon as follows:

- A human figure portrayed on a single or double triangle (the apex of either triangles adhering to each other)
- A pregnant women bulges to the right
- A dead man is turned on his side
- A house is indicated by rows of thatching
- The usual medley of chaprasis, animal pets and distinguished visitors shows how important the icon is
- The sun is represented by three concentric circles round

a spherical object radiating rays

- The moon looks like a banana enclosed within an outline
- A Banyan tree with honeycombs hanging from the branches and monkeys coming to steal the honey and squirrels climbing
- A chair on which a person sits is regarded as Ilda or Rauda
- A mass of people and a dance party are representing the ancestral Saora
- Various animals like lizard, cargaland monitor, porcupine, pangolin, horse, elephant, etc., have a specific relation with the beings represented in the symbols

Table 3 gives the meaning of the items shown in an icon. The pictures in the Table and their position in the icon give an idea of the Saora world-view of plants, animals, birds, insects, spirits, etc.

Among the different types of beings, the ancestral beings are given highest importance. It is because such beings are destructive causal agents of the Saora's illness, sufferings and even death. They ruin the good life of the Saora in many ways: failure of a crop, decrease in secretion of sago palm sap, sudden stop of fruiting in fruit bearing trees and many types of diseases are caused by those ancestral beings. To get rid of the anger of ancestral spirits or as a measure to stop further damage by the ancestral beings, an icon is painted by direction from those ancestral spirits who have caused the problem.

The Saora are conscious about these occurrences. So they take remedial measures not to let or permit an ancestral being to cause any harm. I confine myself to an example of the Saora belief with regard to the relation of plants and ancestral beings. A person after death does not separate itself from his/her kin, but cares for the crops if it is properly given offerings. The spirit

Table 3: Items depicted in the icon

(This table of Behuria and Dash (1991) has been slightly modified with some additional information by the author of this paper)

Sl. No.	Item of Icon	Most frequently found		Less frequently found	
		Name	Meaning	Name	Meaning
1.	Image of Deities	Uyungsum	She has given birth to all Saora-being (Lanjia Saora	Other deities like ancestral spirits, Lobosum, Gungisum, Kittung, Karnosum, Jalliyanem. etc.	The reasons are mentioned inthis chapter (about purposes)
		Angaisum (Moon deity)	Friend of the sun deity		
2.	Human Images	Dancing troops servants of Ildasum	Dead ancestral spirits or	King's palace and court playing of musical instruments, men carryingyoke, men preparing liquor, drinking liquor, hunting troop, etc.	Servants and courtiers of Ildasum
		A person sitting on a chair	Ildasum or Raudasum		
		Shaman curing a patient	Shaman himself/herself.		
3.	Animal	Monkeys	The kin of ancestral spirit evolved from Kureitung, symbol of hereditary system	Porcupine	Responsible for watershed
				Frog	management (for Ilda)
		Dog	(Kambutung) symbol of Ratu'dsum	Horse (most frequently found)	Responsible for rain

Table 3: contd . . .

Sl. No.	Item of Icon	Most frequently found		Less frequently found	
		Name	Meaning	Name	Meaning
		Tigers	(Kinlam) Metamorphosis of man to tiger, symbol of manisum	Fish, Crab, etc.	Vehicle o f Ildasum or Raudasum
		Elephants	Vehicle of Ildasum or Raudasum	Lizard, Snake (most frequently found)	Servant o Uyungsum
4.	Birds	Peacock	Symbol of royalty	Hen, cock and others (unknown ones)	Favourite birds of Ildasum
5.	Trees	Banana	Favourite tree of Raudasum also symbol of A'dur (new eating ceremony)	Bees in a tree	Symbol of Lobosum (earth deity)
6.	Vehicles	Bus, Bicycle	Vehicles of spirits who are from abroad	Aeroplane, train, etc.	Vehicles of abroad spirits
7.	Crops	Crops	Bumper harvest	—	—
8.	Weapons	Bow and arrow	Weapons of Ilda & Idaisum	Swords, guns, etc.	Weapons of Ildasum

puts its soul in the crop plants which it had cropped as a living being before death. So the safeguarding or destruction of the crops depends upon the will of the dead's spirit. Further, the spirit watches over all the properties which he/she had. The spirit puts its soul in Sago Palm trees, in Mahua trees, in crop grains and in fruits of trees, which the deceased had planted or owned and in many things with whom he had an attachment during his earthly life. If the spirits are not perfectly appeased, they may cause a decrease in the resource yield. Sometimes when they appear to their living kin, they ask them about the condition of the trees and plants owned by them before death. The spirit puts its soul in the rice pounding hole and in the rice that is pestled in it. A spirit keeps its eye on all its properties and resources in the material world.

Icons are also drawn in fertility rites. A legend from the village Allingal says, that in the beginning the Saora people did not know agriculture. Once a Saora saw some paddy plants growing near a stream. He saw that the birds were eating its grains. He came to know that they were edible. He collected a lot of paddy grains. He pounded them to get the rice grains. But after pounding the paddy, instead of rice grains, surprisingly, rice flour came out. The Saora got confused. He did not know how to grow the crop again. Then Lobosum (the earth spirit) appeared. He asked the Saora to make a paste from rice flour and water and instructed the man to paint an icon. The man did not know how to paint. He simply threw the paste at the walls with honour to Lobosum. The spirit became appeased with this and after that gave rice grains instead of flour.

The representation of beings in the folk art of the Saora gives the symbolic dualism of existence of the spirits. The chief characteristics of the beings of the cosmos may be distinguished into two categories. Here I consider what is above earth as the visible part of the cosmos and the underworld as the invisible part of the cosmos. In the visible part the living beings are physically present and in the invisible part beings are imaginary; they are the ancestral beings.

Table 4: Saora spatial dualism I

Underworld	*Earth*
Ancestral spirit beings	Living beings (Saora)
Non-life	Life
Immaterial	Material
Invisible	Visible
Darkness	Light
Occasionally mortal	Mortal
Death	Life
Moon	Sun
Peace, entertainment	Punishment, suffering
(Reflection of source)	(Source of power)
Female	Male
Low	High
Affinity	Blood
Peripheral	Central

From the above given structural dualism between the underworld and the earth it is remarkable that the moon is placed with the underworld and the sun is placed with the living beings. Another question in this context is: how are sun and moon perceived in the icon of the Saora? Table 5 shows the Saora perception about sun and moon. Sun and moon represent two different worlds and in-between is the Saora living space. The entire Saora philosophy of life and soul is conceptualised in their world-view and cosmology, myth and reality of beings. The icon becomes highly significant here on the consideration that it is not simply a folk art but rather a symbolic representation of social life, philosophy of life, death and rebirth.

Table 5: Saora spatial dualism II

Moon	*Sun*
1. Known as Angaisum, believed to be the highest experience in the underworld	1. Known as Uyungsum, the highest experience
2. Reflection of sun, power assimilated from sun, cool, peace; Vitebsky considers it as a sub-experience of sun	2. All powerful, yet a scorching agent. "Heat, fire, burns and lesions, blacksmithing, dissolution of forms, redness, menstrual blood, formation of foetus, swallowing, abortion...." (Vitebsky: 1993)
3. Spirits of people who died normally go to the moon	3. Spirits of people who died of an experience go to the sun
4. Represented in an icon as a blunt banana	4. Represented in icon as a circle with radiating rays
5. Saora express the soul's movement towards the moon an upward movement, because the moon is believed to be a peaceful journey of souls towards transcendence	5. Saora express the soul's movement towards the sun as a downward movement because the sun is believed to be a trembling agent
6. Spirits are benevolent representing the cool and peaceful character of the moon	6. The spirits died of an experience are mostly malevolent representing some characters of the sun
7. The moon symbolises life in the icon. It stands for a good crop as the moon has a relationship with vegetation.	7. The sun in the centre of an icon is taken as a central experience from whom all the beings get power; the sun at the top of an icon as the highest experience shows an unidirectional flow of power downwards that reaches plants, animals, human beings and ultimately reaches the moon
8. Spirits (ancestral spirits) in the moon's world (underworld) have their own agriculture and produce their own food. They are rich	8. Spirits (Kulbanji, experience spirits) in the sun-world (above the earth) depend on food of the Saora because they are poor
9. Peace, reflection of source	9. Punishment and source of power

References

Behuria, N.K. & J. Dash (1991): Art, Symbolism and Ritual in a Non-literate Society: A Study of the Social Context and Modern Relevance of Saora Icon, In: Dinanath Pathy (ed.): *Tribal Art — Primitivism and Modern Relevance*, Working Artists Association of Orissa.

Elwin, Verrier (1954): *Tribal Art of Middle India*, OUP, Calcutta.

Elwin, Verrier (1955): *The Religion of an Indian Tribe*, OUP, Bombay.

Vitebsky, P. (1993): *Dialogues with the Dead*, Foundation Books, Delhi.

15

Folk Medicines and Plants used by Tribals of Phulbani and Koraput Districts of Orissa and their Future

Kamala Kumari Patnaik

Introduction

THIS paper indicates the traditionally used folk medicinal plants of Phulbani and Koraput district. Both the Kuttia Kondh – a tribe residing amidst the forests of Phulbani district and the Saora —another tribe residing in the forests of Koraput district, have a strong perception that if they satisfy gods and goddesses and evil spirits by their ritual prayers, offerings and sacrifices, through their priests, Jani (shaman) or astrologers, their ailments/ diseases will be cured. They simply worship the entire environment, viz., lithosphere (hills, mountains, stones), hydrosphere (water) and atmosphere (air), symbolising them by different penu (gods) and spirits/ghosts whom they perceive as physically residing in different spheres to look after their well-being. With this strong perception they seldom go to any physicians or medicine men in their neighbourhood for treatment. However, the medicine man and the astrologer know certain curative properties of herbs/plants growing in the tribe's locality. Concerning the above-mentioned tribes of the Kuttia Kondh and Saora, there is not much documentary evidence of their knowledge or usage of such flora for different ailments. They keep their knowledge only by oral transmission from generation to generation. The data collected from the Kuttia Kondh and Saora on the application of the diverse flora for the treatment of the many ailments have been recorded here. The manner in which they use the many plants and plant parts for birth-control, snake-bites, piles, dysentery and diarrhoea, leucoderma, clinical jaundice, loss of hair, malaria and different types of fever, cough, cold, arthritis, diabetes, wound bleeding, nose-bleeding, gum and dental care, bronchial asthma has been described here. The local names of plants (in Oriya) have been collected and after proper identification the botanical name of each plant is given for general understanding. From the angle of scientific research, it has been observed that medicinal curative properties of certain plants used by these tribals have since come to the light. Hence the traditionally based medicinal plants may be useful for mankind if new technology and research can pinpoint the active principles present therein to establish the facts. The traditional knowledge of these tribal folks may be useful to develop herbal gardens and technical research in developing countries like India for the improvement of medical care at low cost.

Historical aspects of Indian folk medicine

India is a veritable emporium of herbs. The medicinal use of plants was known to the inhabitants of India from times immemorial. As per the Indian Materia Medica, a large number of vegetables and different parts and products of flora and fauna were used for the treatment of diseases by the ancient Hindus. Generally the practitioners of various Indian systems of medicine utilised the flora and fauna found in their environment and accepted those which were found useful after trial, as it used to be practised all over the world.

Ayurveda comes from *Atharvaveda*. There are hundred and fourteen hymns describing some diseases and medicinal practices. This was in the era around 2000 BC. The books written on *Ayurveda* were rewritten by Charaka. They are known as *Charaka Samhita*. They have been written in eight volumes having hundred and twenty chapters in Sanskrit. Sushruta wrote in a similar way and his work known as *Sushruta Samhita* has ten thousand verses in six volumes under one hundred and eighty-six chapters. He wrote on surgery, Kashyap wrote on paediatrics and Vatsayana on sexology. Hence, a very rich and old heritage of *Ayurveda* is prescribed and documented in the Hindu Vedas. This was the age when the art of 'pharmacy' was introduced. Thereafter came the 'creative age', when the active principles of medicinal drugs were adopted. According to Charaka, as given in the chapter 'Kalpasophan', the drugs differ with respect to land, season, source, flavour, taste, potency, post digestive effects and specification and also with respect to one's body, morbid tendency, constitution, age vitality, gastric fire, proclivities, homologation and stage of diseases (Charaka, *Rigveda*).

The Vedic period ends around 1800 BC. The classical period of the Indian medicine begins about 600 BC, when the treatment of various types of ailments were indicated. Due to the Industrial Revolution and the general economic growth in Western countries, there was a systematic growth of Western medicines. Gradually people started to develop confidence in Allopathic medicines, and Ayurvedic medicines were neglected.

Many Allopathic medicines established as life-saving drugs were later found to be harmful to humans. Such medicines proved to have side-effects which cause damage to health. The modern man of scientific aptitude again tries to rotate the wheel of history in search of an alternative source of truth from natural resources. Against this background many attempts have been made in India and abroad to unravel the truth, described in the Vedas with regard to medicinal effects of flora and fauna. Although the Allopathic medicines are found to dominate over the Ayurvedic medicines, many rural and some few remote areas of tribals still continue the traditional practice of using medicines from their natural environment.

Accepting those ideas, the famous Baidya Jhandu started a small factory (Rasashala) in Jamnagar in 1864 with the help of the then king of 'Jamnagar'. Then in 1910 the Jhandu Pharmaceutical work was established in Bombay to manufacture

Ayurvedic medicines on a large scale. Thereafter many small factories were started in different parts of India.

Traditional folk medicinal plants of Phulbani and Koraput districts

Tribal societies residing amidst the forest are basically godfearing. They have a strong perception that if they satisfy their gods and goddesses with prayers and offerings, they can lead their life peacefully. They feel that unless they satisfy them at times with special offerings and prayers they may face difficulties. If they suffer from any diseases or ailments, they immediately call for their priests, Jani or astrologers, to ascertain the cause of suffering. Accordingly they go for the specific traditional prayers, rituals and offerings needed. They seldom go to medicinal practitioners/doctors to get rid of their ailments. Hence, in a man-forest ecosystem they simply worship the entire environment, as different types of penu (gods) and spirits/ghosts residing physically in different spheres. They also do prayers to the sun as the source of energy. The Kuttia Kondh traditionally perceive that the sun dispersed light to the earth when they originally emerged at 'Sapangada' from under the earth (where it was completely dark) to above the earth which was also dark then. The folklore, myths, legends and traditional gaanis (songs) of the Kuttia Kondh reflect their perception of the phenomenon of nature and its relation to man. They follow these rigid principles from generation to generation.

Both the Kuttia Kondh and the Saora go for medicinal plants for their treatment at times. The medicine men know certain curative properties of plants that grow in the surroundings. But there is no documentary evidence about the use of medical plants among them; medicinal knowledge is transmitted by oral tradition from generation to generation. It is their family's property. Sometimes some persons among the tribal folks also develop interest about such medicinal plants and in due course of time they become experts on the subject. At times, it is difficult to get the detailed information about medicinal plants from them due to their conservative attitude. They claim that if they disclose the use of such medicinal plants to outsiders, the medicinal effect of such plants will vanish. However, educated people living in tribal areas venture to acquire the knowledge of their family and diffuse it to their neighbouring localities. When the source has been developed and people are conscious to ascertain alternative medicines, the scientists are eager to explore the possibilities of getting traditional ideas about medicinal plants from the remote areas. In the present investigation, attempts have been made to get information regarding the medicinal plants used by the tribes. Data of medicinal plants have been collected mostly from tribal people. Simultaneously certain information about the mode of use of these plants by the tribes have been recorded. The survey of the medicinal flora has been conducted in certain localities of the Phulbani (Kuttia Kondh's residence) and Koraput (Saora's residence) districts of Orissa. The following findings are excerpts of this survey documenting indigenous medicinal knowledge and practices.

BIRTH-CONTROL

(a) *Ricinus communis*, L. (Oriya: Jada/Gaba/Eranda): One mature seed is collected and after the removal of the testa (seed coat) it is given to the woman on the fourth or fifth day of menstruation. This provides safe contraception for a period of one year. Similarly two seeds taken at a time, provide contraception for two consecutive years. If four-six seeds are taken at a time, the woman acquires permanent sterility.

(b) *Abrus precatorius*, L. (Oriya: Kainch/Gunja): When one white-coloured seed of this plant is taken by a woman along with molasses on the third day of menstruation, she does not conceive for one year. Similarly two seeds taken at a time can check conception for two consecutive years. If three seeds of the white/red variety are taken with molasses in the above manner at a time, the woman acquires complete sterility. However, it is said that the white variety is less harmful (poisonous) than the red seeds.

(c) *Toddalia asiatica*, Lamk. Var. Gracilis, Syn: T.aculeata (Oriya:Tundapoda): When the root is tied to the left arm of a woman, fertilisation is not possible. The bark of the root has got an anti-periodic effect.

(d) *Anona squamosa*, L. (Oriya: Neua) and *Plumbago zeylanica*, L. (Oriya: Chitaparu, Rakta Chitaparu): When equal proportions of root bark of *A. squamosa* and *P. zeylanica* are made into a paste and one tola (= one gram) of this paste is taken orally, abortion takes place within 24 hours.

(e) *Plumeria rubra*, L., Syn : *P. acutifolia*, Poir (Oriya: Kathachampa): When the blunt end of a stem is introduced into the uterus, abortion takes place.

(f) *Ananas sativus*, Lindl. (Oriya: Sapuri): By administering the leaf juice or the juice of a young inflorescence, abortion takes place.

(g) *Carica Papaya*, L. (Oriya: Amruta Bhanda): The milky juice is applied to the uterus to induce abortion.

SNAKE-BITE

(a) *Murraya Koenigii*, Spreng (Oriya: Verusunga): The decoction of the leaf is taken as an antidote for snake venom.

(b) *Rauwolfia serpentina*, Benth. (Oriya: Patala Garuda): The paste of the root is orally administered as treatment against snake-bite and scorpion sting.

(c) *Calotropis procera*, L. (Oriya: Arakha): Two-three leaves are chewed immediately after the snake-bite. This treatment yields a positive result.

(d) *Crotalaria trifoliata*, L. (Oriya: Jhumka): The paste of the dried seed applied with water over the affected area gives relief to the patient.

PILES

(a) *Achyranthes aspera*, L. (Oriya: Apamaranga): One tola of the root along with one tola of *Piper nigrum, L.* (Oriya: Golamaricha), is ground into a paste with water (which earlier is kept for one day inside an earthen pot). This potion taken on an empty stomach continually for three to four days quarterly in a year yields positive result.

(b) *Aloe vera*, L. N. Burman (Oriya: Ghee Kuanri): The juice of one leaf taken on an empty stomach for seven days gives a positive result.

(c) *Euphorbia hirta*, L. (Oriya: Chita Kuti): One tola of mature leaves of this plant and one tola of Jaitri taken by the patient on an empty stomach successively for seven days yields an effective result.

(d) *Hibiscus rosa sinensis*, L. (Oriya: Mandara): Five grams of flower buds mixed with unboiled rice water (Chauladhua pani) are given to the patient to cure the ailment.

DYSENTERY & DIARRHOEA

(a) *Aloe vera*, L. N. Burman (Oriya: Ghee Kuanri): A single flower (fresh or dry) is mixed with a tea-spoonful of milk and taken thrice a day to cure blood dysentery.

(b) *Bryophyllum calycinum salis* (Oriya: Amarapoi/Hema Kedkari): One leaf is grinded with 2½ black pepper and taken along with sherbet (a sweet drink) thrice a day to yield a positive result.

(c) *Coleus aromaticus, Roxb. (Oriya: Rukmini Hatapacha)*: Freshly collected leaves (5 nos.) grinded with black pepper (21 seeds) taken with sherbet (a sweet drink) thrice a day for 3 days give positive result.

LEUCODERMA

(a) *Azadirachta indica*, Juss. (Oriya: Neem): The powder of the wood of an old tree trunk mixed with water to prepare a paste and applied on the affected area continuously for one month, restores a normal colouration to the skin.

A compound leaf of this plant taken with seven seeds of *Piper nigrum*, L. (Oriya: Gola Maricha) in empty stomach for seven days quarterly yields a good result within a year.

(b) *Achyranthes aspera*, L. (Oriya: Sweta Apamaranga): The leaf of this plant mixed with common salt and the juice applied to the affected area for seven weeks gives a positive result.

(c) *Ocimum sanctum*, L. (Oriya: Tulasi): The juice of the leaf along with common salt is rubbed softly over the affected area for three-four weeks.

The leaves when taken orally with 21½ black pepper for one full moon period yield a good result.

(d) *Acalypha indica*, L. (Oriya: Indra Morisha): The juice of the leaf rubbed softly with common salt over the affected area for seven weeks in the beginning state, restores the normal colouration to the skin.

CLINICAL JAUNDICE

(a) *Aloe vera*, L. (Oriya: Ghee Kuanri): The freshly collected leaf juice (one ounce) taken orally for seven days on an empty stomach, yields a positive result.

(b) *Abutilon indicum*, L. Sweet. (Oriya: Pedipedica): Five-seven fresh leaves along with 21 seeds of black pepper (*Piper nigrum*) taken with sherbet give a positive result.

LOSS OF HAIR

(a) *Dendrocinde sinuata* (Blume) Chew. Syn.: *Urtiya sinuata*, B.L. (Oriya: Lata Bichhuati): The unripe fruits along with the stinging hairs of this plant are touched on a thin film of castor oil on the affected area. This mode of application initiates hair growth on the bald area.

(b) *Eclipta alba*, L. Hassk. (Oriya: Kesh Dura): The juice of the leaf is massaged on the affected area daily for three-four months continuously. Hair growth is initiated on the affected area and further loss of hair is checked.

(c) *Weddelia calendulacea*, Less. (Bhringa Raja): The oil of the seeds is applied on the head regularly at bed time. This checks loss of hair and provides the hair with a deep black lustre.

(d) *Musa paradisiaca*, L. (Oriya: Kadali): The leaf ash is mixed with gruel and massaged on the affected area for 10-15 minutes everyday, followed by a bath. This treatment continues for one month to eradicate dandruff.

MALARIA AND HIGH FEVER

(a) *Nyctanthes arbor tristis*, L. (Oriya: Gangasiuli/ Sephali): A paste is prepared of seven freshly collected leaves and 21 seeds of black pepper (*Piper nigrum*) with water and taken on an empty stomach for seven days to produce an effective result.

(b) *Blumea lacera*, DC. (Oriya : Poka Sunga): One tea-spoonful of freshly collected leaf juice taken on an empty stomach for seven days.

(c) *Oldenlandia corymbosa*, L. (Oriya: Gharapodia/Kheta Papada): One tea-spoonful of juice from the entire plant taken thrice a day on an empty stomach for a fortnight yields an effective result.

COLD AND COUGH

(a) *Ocimum sanctum*, L. (Oriya: Tulasi): The leaves of *Ocimum sanctum*, L., *Zingiber officinale*, Rosc. (Oriya: Ada), *Coleus aromaticus*, Roxb. (Rukuna Hata Pochha) leaves, *Piper nigrum*, L. (Oriya: Gola Maricha) seeds, *Achorus calamus* (Oriya: Bacha), *Cinnamomus tamala* (Oriya: Tejpatra leaf), *Curcuma longa*, L. (Oriya: Haladi) rhizome, *Carum copticum*, Ben (Oriya: Juani) seeds and *Cymbopogon nardus*, (Oriya: Dhanwantari) leaves are taken in equal proportion (four-five grams each) and a decoction is prepared. Half cup of this decoction is taken thrice daily to give relief to the patient.

(b) *Ficus glomerata*, Roxb. (Oriya: Yagyna Dimiri): Unripe fruits are fried in ghee and taken for seven days to yield a positive result.

(c) *Solanum khasianum*, Clarke. (Oriya: Bheji Baigana): Ten-fifteen fruits are fried in goat ghee and taken for 7 days to give a good result.

(d) *Aloe vera*, L. (Oriya: Ghee Kuanri): The entire inflorescence fried with ghee gives positive effect. One flower per day for a fortnight also gives a positive result.

ARTHRITIS

(a) *Mangifera indica*, L. (Oriya: Amba): Two-four dried leaves are boiled in 1 litre of water in an earthen pot, until the half is left. This portion is given once a day. This mode of application must continue for one full moon period to give enough relief to the patient.

(b) *Alstonia scholaris*, Brown. (Oriya: Chhatiana): The milky latex is spread over bread and is taken on an empty stomach for one full moon period.

(c) *Tamarindus indica*, L. (Oriya: Tentuli): The paste of freshly collected leaves is mixed with washing soda. This mixture is warmed and applied on the affected area till the paste dries up. This treatment continued for a month with a three months' interval during one year gives a good result.

DIABETES

(a) *Trigonella foenum graecum*, L. (Oriya: Methi): Two-three tea-spoons of dry seed powder or 2 grams of water soaked seeds are taken on an empty stomach daily before breakfast. During this period the diet should contain four-five pieces of garlic and four-five grams of germinated green gram (hot chilli) seeds. This treatment can lower the blood sugar level. The diabetic patient must continue this at least for one year. In acute cases, he must continue throughout life.

(b) *Dolichos biflorus*, L. (Oriya: Kolatha): Ten-fifteen grams of seeds are soaked with water and this water is taken on an empty stomach everyday to control blood sugar level.

(c) *Eugenia jambolana*, Lamk. (Oriya: Jamu Koli): One cup of ripened fruit juice on one tea-spoonful of dried seed powder is taken on an empty stomach before breakfast to decrease blood sugar level.

(d) *Ficus glomerata*, Roxb. (Oriya : Yagyna Dimiri): The juice from a clean root is collected overnight in an earthen pot by making a bore in it. Taken on an empty stomach it lowers the blood sugar level.

WOUND HEALING

(a) *Desmodium triflorum*, D.C. (Oriya: Krushna Parni): The leaf juice is applied to the freshly cut area to check bleeding.

(b) *Kalanchoe laciniate*, L. Personn (Oriya: Patra Gaja): The leaf juice is applied on the wound to check bleeding and it facilitates the healing processes.

(c) *Chrysanthemum* sps. (Oriya: Sebati): The fresh leaf juice is applied on the wound to check bleeding.

(d) *Tritax procumbens*, L. (Oriya: Bisalyakarani): The bruised leaf juice is applied on the wound to cure it.

(e) *Cynodon dactylon* (Oriya: Duba): The paste of the freshly collected plant is applied on the wound to stop bleeding and heal the wound.

(f) *Blumea lacera*, D.C. (Oriya: Poka Sunga): The powder of the dried flower is sprinkled on the wound to stop bleeding.

NOSE-BLEEDING

(a) *Cynodon dactylon* (Oriya: Duba): Two-three drops of the fresh leaf juice is dropped into the nostrils and the patient is to lie on the bed for half an hour.

(b) *Punica Granatum*, L. (Oriya: Dalimba): Dropping two-three drops of flower bud juice into the nostrils checks bleeding.

GUM AND DENTAL CARE

The cut twigs of the following plants are used to brush the teeth:

(a) *Streblus asper*, Lour. (Oriya: Sahada).

(b) *Smilax macrophylla*, Roxb. (Oriya: Muturi).

(c) *Achyranthes aspera*, L. (Oriya: Apamaranga): The root is relatively more effective.

(d) *Mimusops elengi*, L. (Oriya: Baula).

(e) *Pongamia pinnata*, L. Pierre (Oriya: Karanja).

(f) *Azadirachta indica*, Juss. (Oriya: Neem).

(g) *Psidium guajava*, L. (Oriya: Pijuli).

(h) *Eugenia jambolana*, Lamk (Oriya: Jamukoli).

(i) *Bambusa* sps. (Oriya: Baunsha).

BRONCHIAL ASTHMA

Tylophora indica, N.Burman Merrill, (Oriya: Antamula): Of about 129 species of the genus *Tylophora* reported all over the world 28 species are reported in Andhra and six species are reported in Bihar and Orissa. Natural hybrids of this plant are found to be distributed in certain places. This plant exhibits certain peculiarities. Polymorphism of the leaves is indicated in this plant. It is a twinner and a sciophyte. The tribals of Orissa use raw leaves to cure asthma. The following results are obtained from our 18 years of experimentation and time to time consultation with the physicians of S.C.B. Medical College. Similar work was, however, taken up in the Patel Chest Institute with indoor patients.

1. When two leaves are taken in empty stomach daily before sunrise or after sunset for seven days, consecutively relief is felt by the patient.
2. To get better results, this must be repeated quarterly in a year.
3. Fresh leaves in alcoholic extracts (extract in brandy) are taken twice a day, to produce significant effects.
4. Many alkaloids have been separated chromatographically from this plant.

FUTURE

The Council of Scientific and Industrial Research, New Delhi, the Regional Research Laboratory, Bhubaneswar and many other institutes of our country are doing creditable work to isolate the composition of medicinal plants. Many useful compounds have been isolated from different medicinal flora and their efficiency has been established scientifically. A good number of Allopathic medicines prepared on the basis of herbal cures are in the market. Still the country has to find preventive and curative medicines for cancer, AIDS, thalasemia and other such deadly diseases. The recent International Conference on Cancer held at New Delhi came out with a suggestion that certain flora could be explored for the possibility of treating cancer without side-effects. When we are starting to realise the importance of flora for a healthy living and scientists are in search of drugs from the vegetation to eradicate diseases, deforestation is rampant in our country. Our environment has been contaminated due to various types of pollutants from industries and other anthropogenic sources. The tribal who is considered to be the *Man in the Forest* has to be recognised as a bearer of knowledge in folk medicinal plants. After ascertaining the utility of medicinal plants they are to be protected, propagated both by the government and the public. Scientific experiments are to be carried out to establish the active principles behind the herbal drugs. Herbal gardens are to be instituted at different places by adopting modern biotechnological methods. There is a need to make joint efforts by the researchers from the fields of the Allopathic, Ayurvedic and Homeopathic medicine, as a step towards 'the eradication of diseases'. Such an interdisciplinary work on medicinal plants may throw a new light for achieving success, in finding out medicines without side-effects at low cost.

Conclusion

Every herb, potentially, may be of medicinal value. We are still ignorant of many such species and are yet to include them in the list of medicinal plants. Time is ripe to concentrate our heart and mind on identifying the medicinal plants by turning the pages of Vedas, Samhitas, and other pre-historic and historic documentations on medicinal flora. Simultaneously, the present practices involving medicinal herbs which are still prevailing among the tribal inhabitants of forests as well as rural dwellers should not be ignored. Only after achieving such knowledge we can look forward to a better future.

16

Trees for Timber and Trees for the Forest

Aspects of Indigenous Knowledge of Wood and Stone in Karnataka, India

Jan Brouwer

The world of art and craft

AMONG the artisans of the Indian subcontinent, those belonging to the Vishvakarma caste stand out owing to the way they view their own position in the world, the relationship between themselves and between self and society. Their perception and conceptualisation of the world have been worked out in minute detail in their oral tradition, published texts and the tools and workshops.[1]

The Vishvakarma of Karnataka comprise blacksmiths, carpenters, coppersmiths, sculptors and goldsmiths, grouped in a single caste named after the mythological architect Vishvakarma. They are distributed over various sub-castes, so

---

1. The material used in this paper forms part of a larger corpus of data collected in India between 1978 and 1993. The data on the classifications were collected as part of a research founded by the Netherlands Foundation for the Advancement of Tropical Research (WOTRO), The Hague. The data on the crafts lexicon were collected with a grant from the Indira Gandhi National Centre for the Arts, New Delhi.

that no sub-caste has the monopoly over one craft.[2] In this paper, I focus on the carpenters and sculptors, because the procurement of their principal raw material takes them to the forest unlike the three smiths. The metals are invariably associated with female deities, while wood and stone are elaborately classified according to both the genders. These classifications do not only show a meticulous religious or metaphysical interpretation, but also a careful treatment of the environment, as well as an expression of social relationships between the carpenter and the sculptor on the one hand, and between them and the society on the other.

Stones

All Vishvakarma sculptors (*shilpi*) told me that they make use of two sets of classification of stones. All sculptors call the stones they use for idol-carving blackstone (*krishnashila*). However, this blackstone may be granite or a soapstone. The harder the stone, the higher the prestige of the artist.

Having received the order for an idol, the sculptor keeps the dimensions and the shape of the image in his mind and leaves for the quarry to collect the proper stone. Where the order is for an idol of an important temple, he carefully selects the so-called old blackstone (*vriddhashila*). Young blackstone (*balashila*) is considered unsuitable for carving. The criteria for this distinction are vague and seem to be based only on colour: the darker stone is the older one.

When a particular section of the quarry is selected as containing old stone, the sculptor considers its shape and sound. When a male deity is to be carved a male stone (*pumshila*) has to be selected: for a female deity, a female stone (*strishila*). The male stone gives a bass tone when hit with the chisel, the female stone a treble tone. Blackstone, which does not give a sound, is considered as neutral (*napumsakashila*) and is used for pedestals of idols.

2. For a complete ethnography of the Vishvakarmas of Karnataka, including their crafts and world-view, see Brower (1995): *The Makers of the World*.

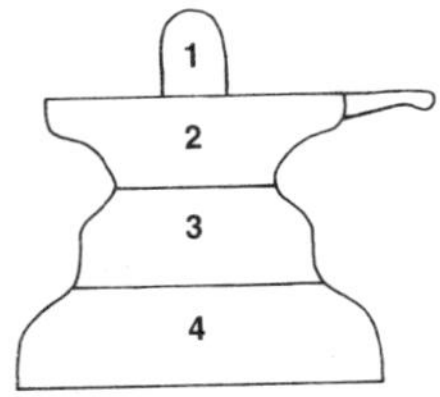

1 = *Linga*; phallic object, Shiva principle.

2 = *Yoni*[3]: vulva, denoting Shakti[4].

3 = upper part of pedestal

4 = lower part of pedestal

Fig. 1: The idol of *Shivalinga*

The choice of stone for four images, however, arrests our attention. Firstly, when the sculptor has to carve an image of *Shivalinga*, the phallic image of Shiva, he requires all the three types of stone. The *linga* proper is made of male stone (*pumshila*). The receiving top of the idol (*yoni*) is made of the same stone, but is referred to as the god Maheshvara. The pedestal consists of two different parts. The upper part is made of female stone (*strishila*), but is called Vishnu who is supposed to include his Shakti (the female Sri[5]). The lower part of the pedestal is made of neutral stone (*napumsakashila*) and is called Brahma. According to the sculptors, the idol of *Shivalinga* also represents the Hindu Trinity, i.e., the deities Brahma, Vishnu and Maheshvara, (Shiva). However, they were not able to explain why the part representing Maheshvara (male deity) made of a male stone, is called *yoni* (vulva denoting Shakti). Secondly, the image of Vishnu, which includes his Shakti (female), is always made of a male stone. Thirdly, the image of Ardhanarishvara, with the right half as Shiva and the left half as his consort Parvati, is to be made out of a single stone, having both (male and female) qualities in the correct place. However, the sculptors add that this particular, not to say peculiar, stone is only available in Kashi (Banaras). Fourthly, if the idol of a village goddess (*Gramadevata*) is made of stone, the image consists only of a carved head made out of neutral stone, which, in this context, is either called *napumsakashila* or *brahmashila*.

3. The external female genitals regarded as the symbol of Shakti.

4. Hinduism: 1. the female principle or organ of generative power/ 2. the wife of a deity, esp. of Shiva.

5. Hindu title of address prefixed to the name of a deity, holy person, etc.

This knowledge, like all the knowledge and skill of stone carving, is transferred orally from father to son. The making of a stone idol depends largely on the skill of the sculptor's fingers, the power of his imagination and his concentration, although he may be aware of the classical (Sanskrit) texts concerning the (sixty-four) arts: the Shilpa-shastra. Many of these, such as Manasara, Mayamata, Kashyapa, Karmakotha, Mulustambha, Silsamgraha, give a classification of stones, but most of the mastercraftsmen do not know them. A few could recite a limited number of hymns (*shloka* in Sanskrit) concerning the name of an image or its form. The hymns on the classification of stones, mentioning colours, shades, odour, spots or dots, blemishes, sound and shape, are not known by the sculptor.

The selection of the proper stone is made by the sculptor himself, who, for this purpose, leaves his workshop for the quarry. After the selection of the stone, a rectangular block of required height and size is cut from the rock by the carpenter (contractor of the quarry) and by his helpers of lower or untouchable castes. The sculptor does not touch the rocks of the quarry and only gives instructions to the carpenter. One similarity between the sculptor and carpenter is that both actually have to cross the border of the village (*grama*) to collect their raw materials from the forest (*aranya*). These two zones are distinguished by religious and social significance rather than by material traits.[6] However, there is one significant difference between the sculptor and the carpenter. The carpenter himself is active in the forest: it is here that he cuts the stones for the sculptors and the trees for timber.

Wood

The Vishvakarma carpenters usually have a choice of ten different kinds of wood for the most common of their products. They obtain the woods from Muslims who own timber depots, or alternatively the patrons provide them with the required wood.

---

6. See Malamoud (1976: 3-20) for an elaborate discussion on this distinction and its characteristics.

However, when a deity or temple chariot is to be carved the carpenter himself, accompanied by the patron, has to go to the forest to select the trees. While cutting the tree the carpenter[7] is identified with the celestial architect Vishvakarma.

The majority of carpenters manufacture agricultural implements, bullock carts, houses, and tools for artisans. The wood they use are consciously classified into two categories:

(i) Those that yield a milky juice or gum (*raktabhutale*), and

(ii) Those that do not yield a milky juice or gum (*nagabhutale*).

All other trees are called 'black trees' (*kappabhutale*). Interestingly the word *rakta* means 'blood', the word *naga* means 'snake' and the word *kappa* means 'black'; *bhutale* literally means 'of the spirit's'.

Table 1: Trees that yield a milky juice or gum

Names of the trees			
Kannada	***Botanical name***	***English***	***Exudation***
Alamara	*Ficus bengalensis*	Banyan tree	milky juice
Bannimara	*Acacia ferruginea*	—	red gum
Halasinamara	*Artocarpus integrifolia*	Jack fruit	dark gum
Bevumara	*Melia azadirachta*	Neem tree	amber
Honnemara	*Pterocarpus marsupium*	Indian Kino	red gum
Gobli, Jali	*Acacia arabica*	Indian gum	red gum
Aralamara	*Ficus religiosa*	Pipal tree	milky juice

Out of the seven trees mentioned, two are not used by the carpenters. The Banyan tree is not used, because it is identified

---

7. In this particular capacity the carpenter is called sculptor (*shilpi*). otherwise he is referred to as carpenter (*badiger*).

with Brahma and is generally regarded to be the male of the Pipal tree. The latter, being viewed as the female to the Banyan tree, is believed to be inhabited by Brahma, Vishnu and Shiva. But in relation to the Neem tree, the Pipal tree is male. The Neem tree is always referred to by the name of the local mother goddess, e.g., Dyamavva, Yellamma, on occasions dressed in a *sari*. The Bannimara (*sami* in Sanskrit) is invariably associated with the mother goddess, particularly with Chamundi. The other five trees are used by the carpenter for the products given in Table 2.

The carpenters apply the aforementioned timbers in the manufacturing of the products mentioned in Table 4.

Furthermore, two of the finished products made of these timbers are associated, or even identified, with goddesses. The carpenters who make the *vina* regard this instrument as equal to Sarasvati,

Table 2 : Five kinds of timber and their use

Name of the trees in Kannada	***Applications***
Bannimara	images of the mother goddess
Halasinamara	vina[8], bellows
Bevumara	palanquin of deities[9]
Honnemara	panels of doors
Gobli, Jali	queer shaft of bullock cart; nave of wheel of bullock cart; shareholder of plough

8. An ancient Indian musical instrument (lute), of which the resonance box is made of the wood of the Jackfruit tree. In parts of Karnataka, Sarasvati is also regarded as the goddess who presides over the sixty-four arts.

9. A passenger conveyance, usually for one person, consisting of a covered or boxlike litter carried by means of poles resting on the shoulders of several men.

Table 3: Trees that do not exude

Names of the trees		
Kannada	***Botanical name***	***English***
Biti	*Dalbergia latifolia*	Blackwood, Rosewood
Matti	*Terminialia tomentosa*	—
Nandimara	*Lagerstroemia lanceolata*	—
Sagavani	*Tectona grandis*	Teak wood

Table 4: Four timbers and their use

Names of the trees in Kannada	***Applications***
Biti	spokes of wheels; door-posts; rods; handles
Matti	central pole of bullock cart
Nandimara	stakes and beams for roof construction
Sagavani	spokes or wheels; door-posts; rods; handles

the goddess of learning. The bellows used by the blacksmiths consist of two or three planks and two or three wooden rims finished with black bullock's leather. Generally, all blacksmiths worship their bellows as the goddess Kali, though some of them regard the bellows as Kali's lion vehicle (*simhavahana*).[10] The name of the tree — Nandimara — is a reference to the bull Nandi, the vehicle of the male deity Shiva.

Besides being used for the carving of images of the goddesses, the exuding kinds of timber are used for products that (a) have a flat surface, e.g., planks of the bellows; palanquin (flat seat) of the deity; or (b) are hollow, e.g., resonance-box of

10. See Brouwer (1995): *The Makers of the World*.

the *vina*; the queer shaft of the bullock cart; the nave and fellows of the wheel of the bullock cart (both the queer shaft and the nave hold the iron axle; the nave and the fellows hold the wooden spokes); the shareholder of the plough. Timbers that do not yield any secretion are used for phallic-shaped products: spokes, posts, rods, handles, long supports, stakes, beams.

Wood and stone in context and text

In the Shilpashastra, particularly in the *Mayamata*, three types of trees are described.

> A straight tree, which is cylindric from its roots to its tip and which has many branches, is called the male tree. A tree which is large at the lower part of the trunk and narrow at the top is called the female tree. A tree which is narrow at the lower part of the trunk and large at the top is a neutral tree.[11]

Considering the classification of timber provided by the Visvakarma carpenters, it can be observed that (a) the timber that yield a secretion (*raktabhutale*), or the products made thereof, are strongly associated with the female deities. Alternatively, the products are of flat surfaces and container type appliances; and (b) the timbers that do no exude (*nagabhutale*) are not associated with a deity, except Nandimara, which is associated with a male deity. All the products made of these timbers have a phallic shape.

For a bullock cart the carpenters prefer to combine the wood in such a manner that the containing parts are made of gum yielding timber and the inserted parts made of non-gum yielding timbers. They add that this is their ancestral practice and not based on technical necessity. For example, Muslim carpenters frequently make all parts of the bullock cart of one type of wood, which is either Teak wood or Indian Gun tree wood. In short, the carpenters have a male/female classification of wood. Furthermore, in the case of the bullock cart, his traditional way

11. Dagens, 1976: Vol.1 Ch. XIII, p.298.

of joining the timber shows that the carpenter establishes a metaphorically complete universe through a violent merging act. This interpretation calls for further study of the bullock cart, which is in principle[12] not different from the war-chariot, or temple chariot, which represents a microcosmos, consisting of the primordial male and female forces.[13]

Strikingly, the only tree which is consciously regarded as male, the Banyan tree, yields a milky juice. And the only tree — the Pipal tree — that is consciously regarded as the female to the Banyan tree also yields a milky juice and not a strongly coloured gum. In classical Hinduism, the Pipal tree (male) is married to the Bannimara (female). However, the Pipal tree can be both male and female (*Atharvaveda* VI, II, I, and *Rigveda* III, 29, 2ff; V,11,6; VI,48,5). A second marriage mentioned in the texts is that between the Pipal tree (male) and the Neem tree (female). A third marriage is mentioned between the Banyan tree (male) and the Neem tree (female). The differences in sex between the trees are as follows : (1) The Pipal tree is either male or female, but no divine specification is given; (2) The Banyan tree is always male, while this tree may be Brahma, Vishnu, Shiva or Krishna. Both the female trees, the Bannimara and the Neem tree, are not identified with a particular goddess.

My field observations show a different picture, but the marriage of the Pipal tree with the Neem tree is also celebrated at the local level. However, in one region, the Banyan tree (male) is married to the Pipal tree (female). In this case, the male tree is not associated with a particular deity. Though the Pipal tree can be both male and female, this tree is always associated with either Brahma, Vishnu or Shiva. At this level, the Neem tree is

---

12. In K. Gollerahalli (Bellary District), the village deity (*Gramadevata*) Kali is taken out in procession on a bullock cart on the fourth night of her yearly festival. In the evening of the fifth day, she follows the same route, but on a temple chariot (*ratha*). See also Sparreboom, 1985: 69.

13. See Brower (1995b): *The Visvakarma World-View*.

always identified with the local goddess, and the Bannimara, or Sami, is associated with Chamundi.

Thus, at the textual level, the Pipal tree stands out as being male and female at the time of marriage to the Bannimara. The ambiguous nature of the Pipal tree is also observed at the village level. On the textual level, however, there is no divine association for the Pipal tree. Whereas, at the village level, this tree is either associated with the three gods Brahma, Shiva and Vishnu, or with Vishnu alone, or with Shiva alone. The lack of divine associations of the female trees at the textual level may be explained by the territorial aspect of the village goddesses. Such territorial links do not fit the fundamental ideas (of purity) in the classical (puranic) literature. Interestingly, when the Pipal tree is married to the Neem tree, the former is not associated with a particular male deity. Apparently, the textual, as well as the contextual data, show similar structures with a similar open slot substantialised by the Pipal tree.

Raw materials, the sculptor and the carpenter

It may be concluded, firstly, that there are significant differences between the principal raw materials, activities and end products of the carpenter and the sculptor. Although both leave the village (*grama*) for the forest (*aranya*) to select and collect their principal raw material, only the carpenter is actually engaged in the cutting of trees and stones.[14] Wood, in contrast to stone, is a less durable kind of material and more susceptible to cyclic influences. Furthermore, two of the major products of the carpenter, the bullock cart and the temple chariot, are intended to move. It may thus be said that the carpenter is associated with a natural as well as a cultural dynamism. By contrast, the product of the sculptor is intended to be permanent as well as static. The carpenter repeatedly joins male wood (*nagabhutale*) with female

14. The carpenter as the maker of the temple chariot (*rathakara*), and in this capacity called *shilpi*, carves himself the tree in the forest and carves the wooden images of deities as part of the decoration of the chariot.

wood (*raktabhutale*). The *naga* or 'snake' is a reference to Shiva whereas the *rakta*, or 'blood' is a reference to feminity, or perhaps to bloodthirsty goddesses. The use of the word *bhutale* for wood here is a reference to the spirits which are believed to inhabit the trees. The process of manufacturing wooden articles is thus never refrained from a metaphorical cosmic copulation realised by the carpenter. By contrast, the sculptor is less instrumental in the realisation of such a cosmic unity. He only puts the finished male part (*linga*) into the finished female part (*yoni*), which both belong to the image of a single god Shiva. Once this idol is installed in the temple, both the three parts remain loose! Finally, the Vishvakarma carpenters rarely make images of deities whereas the sculptors make images of deities only.

The aforementioned differences between carpentry and sculpture demonstrate a difference in purity between the carpenter and the sculptor and this could be one of the reasons for their difference in ritual ranking in the local caste system. Secondly, the contextual classification of wood and stones show a fundamental ambiguity which may contribute to the so-called ambiguous position of these artisans in the local caste systems. The trees are classified into three categories: (1) sap-giving: (2) not sap-giving and (3) other trees. This corresponds with a distinction in male, female and neutral. However, local narratives mention that the Pipal tree, one of the sap-giving trees and which would have been female, is not used as timber, but is considered to be the Hindu Trinity itself, i.e., the deities Brahma, Shiva and Vishnu. It is also a wood that is either male or female depending on the context in which it appears with other wood. The classification used by the carpenters speaking about 'trees that give a sap', and about *raktabhutale*, suggest a difficulty since the general statement 'sap-giving' may refer to those trees which exude a milky juice, as well as to those that yield a (often red) coloured gum (like blood). Could the alternate sexuality of the same Pipal tree have something to do with an association of the milky juice with either semen or mother's milk? What interests us here is that the sexual ambiguity concerns a tree that is considered to be Brahma, Shiva and Vishnu, i.e., all male deities together. Although the carpenters do classify the ambiguous Pipal tree, they do not use it as timber, as if they want to avoid the ambiguity itself. Or, in other terms, they themselves are perhaps not pure enough to deal with such ambiguous, hence dangerous, material.

Thirdly, stones are also classified into three categories called male, female and neutral. The sculptor selects the stones in such a way that the gender of the stones corresponds with the sex of the deity which he has to make in stone. This correspondence shows an ambiguity. The image of Shiva called *Shivalinga* consists of two complementary principles: the male (*linga*) and the female (*yoni*) respectively, which are both made of male stone. Consciously that part of the image that is called *yoni* is also considered a form of the male deity Maheshvara (another name of Shiva). The *linga* and *yoni* together form the actual image of Shiva. The pedestal of this image consists of two parts. The upper part is Vishnu made of a female stone, because Vishnu implies his consort Shri (*shakti*). However, when a separate image of Vishnu is made, the idol is entirely made of a male stone. Thus, Shiva as well as Vishnu have, each in his own way, a double sexuality. The lower part of the pedestal is called Brahma and made of neutral stone. Separate images of Brahma are never made. But when an image of the (*adi*) *shakti* is made, the neutral stone is used. This suggests a complementarity of Brahma and the *adisakti* or the original power that existed even before the world. In other words, the three supreme male deities are all expressed in terms of a complementarity though in different terms. The more earthly Vishnu (the preserver who incarnates ten times) implies his female consort and appears even as the female deity Mohini being the consort of Shiva; the less earthly Shiva (destroyer and creator; the erotic and ascetic) is represented with a male and female half (*linga* and *yoni*); Ardhanarishvara); the heavenly Brahma, creator of all, has an association, via the imagery of stone, with the original power of the universe. The meaning of the ambiguity in the three expressions is a subject for further study, but it is clear that the sculptor deals with it in his craft itself.

Wood and stone and the crafts lexicon

Neither timber and stones, nor any other raw material used by any of the five Vishvakarma craftsmen appear as a term in their crafts lexicon. This lexicon which is not to be confused with the crafts technical vocabulary covers the following areas: (i) categorical terms; (ii) craft related terms; (iii) terms relating to people; and (iv) terms relating to customers. The lexicon consists mainly of nouns and a few adjectives and verbs. The nouns follow Kannada declensions and the verbs Kannada conjugations.[15]

The lexicon terms for society in general, and for caste names in particular, show two terms of relevance here, namely 'forest' and 'firewood'. Notably, artisan tools, as terms, are absent from these lists. Another list states the lexicon terms for the self (as the Vishvakarama caste). Herein the artisans' tools appear frequently, as well as such terms as 'chip of wood' and 'joint of wood'. This latter observation shows their identification with the crafts. No term of the lexicon, whether for self or society, is related to timber or stones. The lexicon terms 'forest' and 'firewood' appear in the list for society. This suggests that the forest including firewood belongs to society. Concomitantly, however, wood for timber is excluded from society. Equally absent from both the lists (society, self) are terms for animals.

It may thus be concluded that in the perception of the artisans, four different domains are distinguished: (1) society including forest; (2) self including crafts; (3) wood for timber;

---

15. The crafts lexicon is to be seen as a traditional means of protection of the craftsmen and his craft, so as to speak, of a 'copy-right'. In a situation in which the division of labour is directed towards the whole of society, and competition restricted to intra-artisan sub-castes, the crafts lexicon could guarantee safety and security of the crafts workers concerned. In this way, excellence of technology, technological invention and specific skills could be protected and preserved. The lexicon has thus two protective functions: (a) to protect trade secrets, and (b) to protect certain customs and habits.

and (4) fauna. The carpenter's classification of wood into trees for timber and trees for the forest now gains another meaning. The trees for the forest are those for society, while those for timber are for the self. In other words, the carpenter disconnects the self from society and places timber under the self. He thus carves out his social position conforming his ideal of the Brahmanical status, or even a status superior to brahmanas, i.e., one placed outside the social world.

Indigenous knowledge for development

The sculptors' and carpenters' classifications of stones and wood show how important resource persons they are for the knowledge of the forest. Their indigenous knowledge is transferred either from father to son, or from teacher (*guru*) to student (*shishya*). In the newly emerging economic scenario these aspects of indigenous knowledge are rapidly disappearing. The artisans' knowledge of stones and wood is locally no longer considered relevant. But the analysis of their intimate knowledge of these forest materials shows a careful treatment of nature even in historical periods when its produce was not yet scarce. From the comparison of the contextual and the textual data on stones and wood, we learn that only a detailed study of the practitioners is relevant for gaining insight in the selection of nature's products. For the students of Indology it raises the question whether the indigenous practice or the theory was first.

From both the classifications and the terms of the crafts lexicon, it is clear that, unlike in the West, the social, the economical, the technology and other domains are intertwined. In the European view, these domains are strictly separated. In contemporary Europe, the concern for environment (forestry) is not linked with the social and the economical. The ecologists and environmentalists of the West strive to preserve the unspoilt just for itself. In India their ultimate goal is a socially just and ecologically sustainable model of development (see Guha, 1995:20ff). The thrust of the modern and indigenous Indian concern for the environment, in contrast to that of the West, are its questions of priority and distribution within human society. The analysis of the artisans' perceptions of wood and stone thus also demonstrates that the social, religious, ecological and technological domains are intertwined. A modern intervention based on a perception of society divided into separated domains would thus meet serious difficulties, if not resistance. The mapping out of the domains through a study of local perceptions prior to intervention is thus necessary to attain the goal of sustainable development.

References

Brandis, D. (1906): *Indian Trees*, London.

Brouwer, Jan (1991): Die Gesellschaft als Körper: Geschlechtspezifische Arbeitsteilung in Südindien, In: *Zeitschrift für Semiotik*, Band 13, Heft 1-2. pp. 23-42.

Brouwer, Jan (1995a): *The Makers of the World. Castes, Craft and Mind of South Indian Artisans*, New Delhi: Oxford University Press.

Brouwer, Jan (1995b): The Visvakarma World-View, In: *Prakrti — The Integral Vision*, Vol.I: Primal Elements: The Oral Tradition. Kapila Vatsyayan (ed.), New Delhi: Indira Gandhi National Centre for the Arts, and D.K. Printworld.

Dagens, B. (1976): *Mayamata, Traité Sanskrit d'Architecture. Edition critique, traduction, notes*, Pondicherry: Institute français de l'Indologie.

Enthoven, E. (1924): *The Folklore of Bombay*, Oxford: Clarendon Press.

Guha, Ramachandra (1995): Variations of Environmentalism, Paper Presented to the IDPAD Seminar on Comparative Research and Alternatives in Development, Bangalore: 16-19 March.

Gupta, S.M. (1971): *Plant myths and traditions in India*, Leiden: Brill.

Gupta, S.S. (1965): *Tree symbol worship in India*, Bombay: Indian Folklore Society.

Macdonell & Keith (1958): *Vedic index of names and subjects*, London: John Murry & Co.

Malamoud, Ch. (1976): Village et forêt dans l'idéologie de l'Inde brahmanique, In: *Archives Européennes de Sociologie*, Vol. XVII.

Meyer, J.J. (1937): *Trilogie altindischer Mächte und Feste der Vegetation*, 3 Vols. Zürich / Leipzig: M. Niehans Verlag.

Philpot, J.H. (1897): *The sacred tree or the tree in religion and myth*, London: Macmillan.

Sparreboom, M. (1985): *Chariots in the Veda*, Memoirs of the Kern Institute No. 3. Leiden: Brill.

Stutley, M. and Stutley, J. (1977): *A dictionary of Hinduism. Its mythology, folklore and development 1500 BC-AD 1500*, Bombay: Allied Publishers.

Viennot, O. (1954): *Le culte de l'arbre dans l'Inde ancienne*, Paris: Presses universitaires de France.

Watt, G. (1972): *A dictionary of the economic products of India*, 6 Vols. New Delhi: Jayyed Press (Second print).

Index